BIG WORLD

WORLD

ABUNDANCE WITHIN PLANETARY BOUNDARIES

SMALL

PLANET

JOHAN ROCKSTRÖM AND MATTIAS KLUM

Yale UNIVERSITY PRESS

WITH PETER MILLER

New Haven and London

First published 2015 in the United States by Yale University Press
and in Europe by Max Ström.

Yale University Press books may be purchased in quantity for
educational, business, or promotional use. For information, please
e-mail sales.press@yale.edu (U.S. office) or sales@yaleup.co.uk
(U.K. office).

Editor Peter Miller
Illustrations/Graphs Jerker Lokrantz, Azote
Design Patric Leo
Layout Patric Leo, Petra Ahston Inkapööl
Prepress Monika Klum, Fredrika Stelander, and Graphicom
Printed on FSC paper by Graphicom, Italy 2015

Library of Congress Control Number: 2015940944
ISBN 978-0-300-21836-7 (cloth)

A catalogue record for this book is available from the British Library.

10 9 8 7 6 5 4 3 2 1

FSC
www.fsc.org
MIX
Paper from
responsible sources
FSC® C013123

CONTENTS

PREFACE

A PARTNERSHIP FOR CHANGE

IT WAS A BITTERLY cold night in Copenhagen. The big climate summit of 2009 was over and everyone was heading home. At the cavernous Bella Center, where delegates from 192 countries had spent two weeks in contentious negotiations, no one remained but an army of workers rolling up power cords and taking down hardware. The tens of thousands of protestors who had filled the city streets were long gone, along with the police helicopters that had hovered overhead. President Obama and the Americans were gone. Premier Wen Jinbao and the Chinese were gone. The diplomats, journalists, and activists who had come here to witness policymaking history were gone, and—without anything resembling success to point to—most had departed with the same word on their lips: "failure."

"What went wrong?" Mattias asked as we stood outside the Bella Center. "The people I talked to had such high hopes."

Just a few days before, at an exhibition of his photographs, Mattias had given a short presentation to a high-powered group about the global effects of logging in Borneo. Tony Blair, former prime minister of the UK, had been there, as had HRH Crown Princess Victoria of Sweden, and Gro Harlem Brundtland, Special Envoy on Climate Change for the United Nations (UN). Having visited Borneo many times during his nearly three decades as a wildlife photographer and filmmaker, Mattias had spoken from the heart when he described the rampant destruction that has claimed 75 percent of the island's lowland rainforest, threatening both the reclusive Penan tribe and wildlife such as orangutans and pygmy elephants.

"Everybody seemed to get it," Mattias said.

I had a similar experience that same week, when I took part in a European Union (EU)-sponsored side event on climate change. As panelists, we were asked to offer scenarios for stabilizing global temperatures at no more than 2°C (3.6°F) above pre-industrial levels—the target being widely proposed at the conference. I had remarked to the group that a climate deal alone—as difficult as it might be to achieve—might not be enough to meet a 2°C goal, since the world's climate was intertwined with other urgent problems such as biodiversity loss and ocean acidification. But other speakers were less cautious, expressing confidence that the goal was still reachable with existing technologies. The general mood was quite positive.

"I'm not really sure what happened," I replied. "But I think two factors still haunt us. One is a deep distrust between rich and poor. And the other is that we're still blind, despite all the science, to the fact that wealth in the world depends on the health of our planet."

The Copenhagen summit was supposed to be the culmination of a long journey. As the fifteenth session of the UN body charged with carrying out the 1992 climate treaty forged in Rio, and the fifth session of the governing body responsible for the 1997 Kyoto Protocol, it was supposed to be the meeting where the world finally came together to sign a new legally binding agreement on climate change. The Kyoto Protocol was set to expire in 2012.

In anticipation of the Copenhagen meeting, the prestigious scientific journal *Nature* had published an article in its September issue that I'd co-authored with nearly two dozen international researchers. In the article, entitled "A Safe Operating Space for Humanity," we proposed to track "planetary boundaries" for critical natural systems such as the global climate, stratospheric ozone, biodiversity, and ocean acidification. If humanity wanted to avoid triggering potentially disastrous tipping points, such as the melting of polar ice sheets, extreme storms, or mass extinctions of wildlife, we argued, then the world needed to know where the boundaries for such thresholds were located, which meant measuring them and tracking them. More positively, we wrote, by identifying these boundaries, humanity could chart a safe path into the future for generations to come, opening the door for greater prosperity, justice, and technological advancement.

What we presented, based on the latest science, was evidence that the world needs a new paradigm for development, one that pursues alleviation of poverty and economic growth while staying within the safe planetary boundaries that define a stable and resilient planet.

The editors of *Nature* called our proposal "a grand intellectual challenge," saying that it could provide "badly needed information for policymakers," which was exactly what we were hoping for, of course. As expected, the paper stimulated a lot of debate, including criticism of our methods and assumptions. This was science, after all, which lives and breathes skepticism and disagreement. We wanted to challenge our scientific peers.

But new research and scientific debates since publication of the paper have verified the need for planetary boundary thinking. The most recent science confirms that as long as we manage, within safe boundaries, our planet's key systems—the climate system, the stratospheric ozone layer, ocean acidification, the remaining forests on Earth, and secure enough freshwater in our rivers and landscapes, safeguard biodiversity, and avoid air pollution and release of chemical compounds—we stand a good chance to secure a prosperous future for the world for many generations to come.

But that was the scientific side of the coin. We also aimed our proposal at a broader audience, including business leaders and political leaders—like those meeting in Copenhagen. We wanted to give the world a new framework to redefine global development by reconnecting economies and societies to the planet. In so doing, we wanted to create a tool providing a practical and comprehensive way to measure human impacts on Earth, and guide our common endeavor toward a sustainable world development, before it was too late.

It soon became obvious, however, that the time hadn't come yet for such ambitious goals. Despite all the promising talk, world leaders failed miserably in Copenhagen to agree on targets to stay within a safe global budget for carbon in the atmosphere, one of the nine planetary boundaries, by reducing greenhouse gas emissions. Negotiating sessions unraveled as groups of delegates walked out. Unofficial meetings sprang up right and left. When President Obama and the leaders of four other nations announced on the last day that they'd privately reached an accord, the rest of the conference felt excluded. The media called the summit "disappointing" and "a missed opportunity." Andreas Carlgren, Sweden's Environment Minister, went further, describing the talks as a "disaster" and "a great failure." Outside the Bella Center, protestors cut off their hair in frustration. "The city of Copenhagen is a crime scene tonight, with the guilty men and women fleeing to the airport," reported John Sauven, executive director of Greenpeace UK.

If one looked hard enough, of course, one could find crumbs of progress coming out of Copenhagen. It clearly represented the moment, for example, when world leaders finally recognized that climate change wasn't just an environmental issue, but also a social and economic one. That meant that any future climate solution would also require fundamental changes in our economies, financial systems, how we build our cities, produce our food, and relate to one another—which was hardly a small change in thinking. Still, there was much to be depressed about as we stood outside the Bella Center.

TWO APPROACHES TO LIFE
As a photographer and a scientist, respectively, Mattias and I traveled in different worlds. I always thought it was a scientist's job to appeal to the

rationality of others. But it was becoming painfully clear to me how naïve it was to assume that, just because the facts were on the table, people would make the right decisions. That wasn't the way the world worked. Any profound changes in society would only happen, I knew, if a large enough percentage of citizens were convinced, felt engaged, and believed in something. A deep mind-shift was required for genuine change, and that couldn't be reached through numbers alone. It had to come from both the heart and the brain.

As a photographer and filmmaker, Mattias had captivated lecture audiences around the world with his startling images and documentaries about wildlife and native cultures. The stories he'd told about encountering tigers in India's sal forest, or of getting within striking distance of a cobra's fangs for one last shot, had enthralled listeners from Seattle to Stockholm, from Beijing to Rio de Janeiro. Yet he, too, had realized that it wasn't enough to make people care about such places. They also needed evidence-based information to persuade them to take action.

In our different ways, in other words, we'd been moving toward the same conclusion: That the best case for a new relationship with nature would be one that bridged the gap between science and the arts, the rational and the emotional, in the service of change. The knowledge and technology existed, we both believed, to solve the world's many problems, and the future was full of opportunity as well as of danger.

Why not combine our talents and tell that story together?

A NEW NARRATIVE

We did just that in the spring of 2012, launching our first book, *The Human Quest: Prospering Within Planetary Boundaries*, with barely enough time to hand out copies to delegates at the UN Summit on Sustainable Development in Rio de Janeiro. That conference, known as Rio+20, was a follow-up 20 years later to the 1992 meeting in Rio, which was the first to bring together concerns about the environment and social development. With support from the Swedish Post-code Lottery, we presented our book, with a foreword by President Bill Clinton, to more than 130 heads of state and government.

At the same time, we were pleased to notice that researchers and policy-makers were starting to adopt our concept of planetary boundaries as they framed their discussions of climate change and other global issues. The term was embraced by the UN High-Level Panel on Global Sustainability, for example, as well as by organizations such as Oxfam and the World Wildlife Fund, and supported by both the EU and the large non-governmental organizations' (NGO) forum in Rio, as an important new way of framing sustainable development in our globalized world with rising global environmental risks. In fact, it was even included, for a time, in the working documents at the Rio+20 conference itself, which was quite encouraging.

We knew that this doorstopper of a book wouldn't be everybody's cup of tea. Written primarily for the professional crowd, it was packed with footnotes, references, and more than 40 data-heavy charts and diagrams—as well as Mattias's amazing photographs—tipping the scales at more than 2 kg (4.4 pounds). But we did this deliberately. We wanted the book to establish an authoritative baseline for a new dialogue on our novel thinking about human development. In the back of our minds, we also harbored the hope of reaching out to a broader audience, as intelligent people everywhere tuned in to the world's predicament. Heat waves, droughts, floods, and other forms of extreme weather were prompting TV reporters almost weekly to ask if climate change was happening faster than expected. Biologists were warning that habitats for countless species were shifting, threatening extinction for many, risking collapse of ecosystems supporting human wellbeing.

We were also encouraged to observe that business and community leaders were connecting the dots, realizing how changes to the environment were creating a new economic landscape, with both threats and opportunities for their own constituents. While the world's leaders continued to dither, debating "top down" policies on climate change and other urgent issues, the rest of the world was already in motion, working on "bottom up" solutions at home, in communities, in boardrooms, and across digital networks. Mattias and I wanted to reach out to these groups as well, to share our passion, experience, and stories with them, by providing a state-of-the-art scientific analysis, presenting a new paradigm for world development within a stable planet, and marry these with a narrative of beauty and hope.

The world needs a new narrative—a positive story about new opportunities for humanity to thrive on our beautiful planet by using ingenuity, core values, and humanism to become wise stewards of nature and the entire planet. The dominant narrative until now has been about infinite material growth on a finite planet, assuming that Earth and nature have an endless capacity to take abuse without punching back. That narrative held up as long as we inhabited a relatively small world on a relatively big planet—one in which Earth kept forgiving all the insults we threw at her. But that is no longer the case. We left that era 25 years ago. Today we inhabit a big world on a small planet—one so saturated with environmental pressures that it has started to submit its first invoices to the world economy: the rising costs of extreme weather events and the volatility of world food and resource costs.

We need a new way of thinking about our relationship with nature, and how reconnecting with the planet can open up new avenues to world prosperity.

That's why we wrote this book.

We've divided the book into three parts. The first part summarizes the urgent predicament we're facing as Earth responds to massive human impacts. It explains our concept of planetary boundaries and details the major threats to our survival if we ignore them. The second part makes the case for a new way of thinking about prosperity, justice, and happiness on a sustainable planet. We believe that preserving nature's beauty is a universal value among all nations, cultures, and religions. No one wakes up in the morning with the intention of making Earth an uglier place. Whether they express it in scientific, humanistic, or religious terms, all peoples share a deep sense of responsibility for our home. In the third part we offer practical solutions to the biggest challenges facing humanity, such as feeding nine billion people or powering tomorrow's economies.

Although we find the latest data about human impacts on Earth truly alarming, we believe this book tells a positive story that will inspire hope, innovation, and countless new opportunities for wise stewardship of the planet. Told in the language of science and photography, our new narrative is about what matters most to you and me—to sustain the remaining beauty on Earth, not for the sake of the planet (she doesn't care) but for ourselves and future generations, a world that in less than two generations will host nine to ten billion people, all with the same right to a thriving life on Earth.

People all over the world are joining this conversation. On Sunday, September 21, 2014, about 400,000 demonstrators crowded into the heart of Manhattan for the People's Climate March. It was an overwhelming sight. On Central Park West, people of all ages and all walks of life lined the avenue along the length of the park, from Cathedral Parkway in the north to the front of the demonstration at Columbus Circle in the south. This was not a "normal" demonstration. Of course, the die-hard environmentalists were there, from anti-nuclear power activists to the doomsday neo-Malthusians. But there were also middle-class parents with their teenage kids, business leaders and innovators, and in front of them all, Ban Ki-moon, the UN Secretary General himself.

Together with sister marches in many cities around the world, the event was a call for political leadership on climate change on a scale we'd never seen before. It was impossible not to view it as a potentially significant moment, perhaps even a social tipping point. It wasn't just the large numbers of participants. It was also the new narrative being put forth, one of working with the political system to enable a better type of economic growth in the future, steering away from the dirty, costly, and risky road we're on today, and instead choosing a clean, economically and socially attractive, path toward a sustainable future.

The day before, some stunning survey results had been released by the Global Challenges Foundation. The researchers had posed the question of whether humans are the cause behind climate change, and whether political leadership is needed to solve the problem. In countries like Sweden, the UK, and Germany, we know that awareness is very high, with some 70 percent of respondents saying humans are the main cause behind climate change. I often hear people claim that this is an exception and not representative of the world at large, because "you Swedes are so environmentally conscious." But the survey posed the same question to people in China, India, and Brazil. And, to my great surprise, the figures in these countries are even higher than in Sweden!

Sure, one has to be careful in interpreting this kind of survey. But still, results like these point out the great mismatch between citizens' awareness and willingness to see political solutions and the limited media attention and weak political leadership we see today. Combined with the enthusiasm I witnessed at the People's Climate March, they represented a clear and unprecedented message from citizens to their leaders: You have a mandate to act.

In December 2015, world leaders will gather again in Paris for another major summit, the UN Climate Change Conference. Our hope is that, this time, they'll correct what went wrong in Copenhagen and forge a new globally binding climate deal. But we don't think you should have to wait for heads of governments to pronounce acceptable solutions to our planetary crisis. With the information in this book, we urge you to embrace challenges for yourself by engaging with others in conversations, dialogues, and meetings—just as those hundreds of thousands did in New York. By sharing this knowledge, as well as your concerns and dreams about thriving families, communities, businesses, and nations, you can help advance the process of reconnecting humanity with nature in harmony with Earth.

That's our wish for this book—and for you.

OUR TEN
KEY MESSAGES

1.

OPEN YOUR EYES

The numbers are overwhelming. The planet's under unprecedented pressure. Too many forests cut down. Too many fish pulled from the sea. Too many species gone extinct. Earth's being battered by humanity—and it's coming from every direction. Greenhouse gases. Ocean acidification. Chemical pollution. It has all reached a point where our future is at risk. For the first time in human history, we may have pushed the planet too far.

Tebaran, a hunter in Borneo, fears a difficult future for indigenous people as logging operations destroy the rainforest. Deforestation also impacts biodiversity and global climate.

2.

THE CRISIS IS GLOBAL
AND URGENT

It's happened so fast. In just two generations, humanity has overwhelmed Earth's capacity to continue supporting our world in a stable way. We've gone from being a small world on a big planet to a big world on a small planet. Now Earth is responding with environmental shocks to the global economy. This is a great turning point. Our home is changing, and our future depends on what we do next.

Construction projects like this one in Hong Kong are part of a boom in economic growth and population. Two thirds of the cities needed by 2030 have not yet been built.

3.

EVERYTHING IS
HYPER-CONNECTED

In a shrinking world, seemingly unrelated events can be links in the same chain of cause and effect. Nature, politics, and the economy are now interconnected. How a worker commutes in Stockholm affects the farmer in Ecuador. The web of life is fully connected, encompassing all of the planet's ecosystems, and every link of the chain matters.

The aurora borealis plays across the sky over Svalbard, an Arctic archipelago. By reflecting sunlight back into space, the white surface of Arctic ice helps to cool the planet.

4.

EXPECT THE UNEXPECTED

As Earth changes, we can expect surprises. The forces driving planetary change are complex and likely to create sudden, unexpected problems. In the past, we could assume that the big systems we relied on—from political to ecological— were stable and predictable. Increasingly today, and most certainly in the future, the only constant will be change. Surprise is the new normal.

Robert, a boy from Nyungwe, Rwanda, will grow up at a time when protection of ecosystems and human progress will prove equally important.

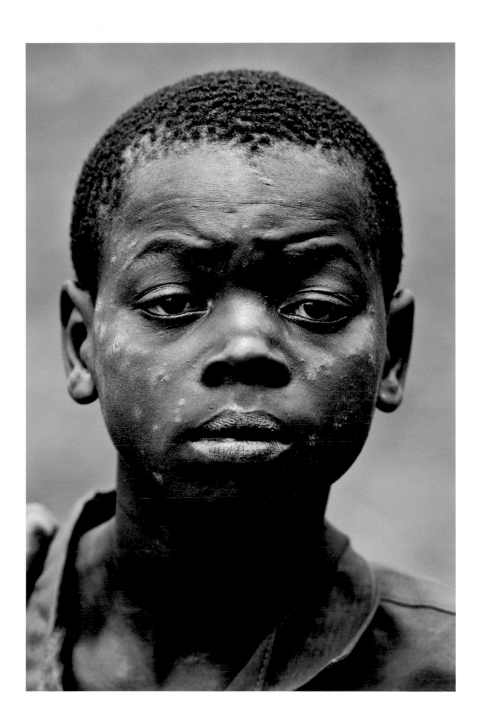

5.

RESPECTING PLANETARY BOUNDARIES

As many scientists have warned, nothing is more important than to avoid triggering disastrous tipping points in Earth's fundamental processes. Fortunately, we now have enough knowledge and data to define planetary boundaries which, if transgressed, could lead to catastrophic problems. If we respect those boundaries, we can follow a safe path to unlimited opportunities into the future.

The rich biodiversity of the Selous Game Reserve in Tanzania is critical for landscape resilience, which contributes to the stability of the Earth system.

6.

THE GLOBAL MIND-SHIFT

Ever since the industrial revolution, we've had this crazy idea that our actions are without consequences. That we can take nature or leave it. But as any farmer can tell you, that isn't the case. It's not a question of choosing jobs or the environment, because they depend on each other. That's why we say we need a mind-shift to reconnect people with nature, societies with the biosphere, the human world with Earth.

Venom from snakes such as this Jameson's mamba from Cameroon is helping researchers develop new drugs to treat heart disease in humans.

7.

PRESERVING THE REMAINING
BEAUTY ON EARTH

We take it for granted, the world that we love—and we're destroying it so quickly. The light of dawn on the prairie. The silvery flash of fish in a stream. The cry of a hawk over a forest. Everybody has their own idea of the beautiful, and we'll surely miss it when it's gone. It's time to fight for the remaining natural systems that support the beauty on Earth—not just for their sake but primarily to safeguard our prosperity.

Lianas twine around dipterocarp trees in Malaysia's Danum Valley Conservation Area. Rainforests provide a vast number of ecosystem services for humanity.

8.

WE CAN TURN THINGS AROUND

We have the tools to do what's required—the intelligence, creativity, and technological know-how. We can reverse the negative trends. We can feed nine billion people without destroying our forests. We can deliver power to our economies without burning fossil fuels. But the only way to achieve prosperity is through green growth. This is not a burden or sacrifice. It is an investment in future world prosperity. Business-as-usual is no longer an option.

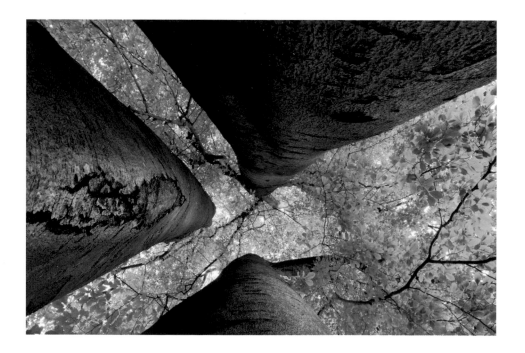

Reconnecting societies with the biosphere is key to soaring opportunities for the future.

9.

UNLEASHING INNOVATION

Humanity has an incredible ability to overcome even the most daunting of challenges. Once people understand the risks of continuing along the current path, they'll search for creative—and profitable—alternatives. That's how innovation works. The planetary boundaries will help. By defining thresholds and a maximum allowable use of resources, ecosystems, and the climate, we can trigger a new wave of sustainable technological inventions thanks to an abundance of ideas and solutions for human prosperity and planetary stability.

This ingenious storm-surge barrier was built to protect Rotterdam Harbor from flooding. By unleashing innovation within planetary boundaries, humanity can make abundant progress.

10.

FIRST THINGS FIRST

Let's be realistic. Inspiring a mind-shift to sustainability could take a generation, and we should have started long ago. If we wait 30 more years, it will be too late. So we advocate a two-track approach: 1) tackle the most urgent problems right now, such as climate change, nitrogen and phosphorus overload, and loss of biodiversity, but also 2) do everything we can to reconnect with nature over the long term. Earth deserves nothing less. Our world depends on nothing less.

A field of rye evokes the ideal of abundant food for all. To achieve that with a booming global population will require a new revolution in agriculture.

SECTION 1

THE GRAND CHALLENGE

HÅKAN NORDKVIST, who manages sustainability innovation at the IKEA Group, told a fascinating story recently. He was taking part in a panel discussion on the stage of the Grand Hotel's elegant Winter Garden in Stockholm. In the audience were 150 business leaders, policymakers, and experts eager to hear the latest thinking about sustainability and business.

As Nordkvist explained, IKEA is now deeply involved in transitioning its energy supply to renewable sources. The company owns about 140 windmills and has installed more than 550,000 solar panels on its buildings around the world. By 2020, IKEA plans to produce more energy than it consumes, contributing to global sustainability while making IKEA more competitive.

What caught my ear was Nordkvist's comments about a discussion in IKEA's boardroom. When asked about the idea of investing in renewable energy, the company's financial people apparently had urged against it, maintaining there was no economic case to do so. Investing in solar and wind would be more expensive than the alternatives, they'd pointed out. But when Ingvar Kamprad, the founder of IKEA, spoke up, he insisted that they should go ahead anyway. When asked why, Kamprad's answer was simple: "Because it is the right thing to do."

The story resonated with me because it reflects a huge shift taking place in the way that businesses treat sustainability. A decade ago, many companies dismissed sustainability as an aside—an engagement in corporate social responsibility (CSR) outside of core business. But today, companies are integrating sustainability into core business strategy. Issues related to climate change or ecosystems are no longer the exclusive realm of directors for the environment; they're an agenda item for boardrooms. Resource efficiency, circular business models, low-carbon value chains, and environmental accounting are all key pieces of strategies, not only to be profitable but also to produce long-lasting companies.

Some game-changers are happening already. The B Team, a non-profit initiative started by Jochen Zeitz of Puma and Richard Branson of Virgin, has been promoting serious integration of sustainability across business sectors. Unilever, Royal DSM, and Walmart are each pioneering sustainable value chains in the food industry. Carl-Henrik Svanberg, chairman of BP, has lent his support to a global price on carbon, saying it's not reasonable that oil companies should be able to pollute the atmosphere for free. General Electric (GE) has saved 300 million US dollars (USD) from energy and water improvements in their own operations (since 2005) and has generated more than 160 billion USD in revenue from technology solutions that have saved money and reduced environmental impacts.

More and more businesses have reached the conclusion that sustainability gives them an edge over competitors who can't keep up with change. The old strategy of exploiting Earth's natural resources and polluting the planet as much as possible no longer works, as *The Economist* noted not long ago in a special issue about the future of oil. Industries that continue to invest in fossil fuels are doomed to fail, the magazine maintained, not on environmental grounds primarily, but because clean energy will become so abundant, cheap, and attractive it will outcompete the volatile, dangerous, unhealthy, risky, and increasingly scarce fossil alternatives.

The writing's on the wall. As the chapters in this section detail, humanity's very survival depends on a deep shift in the way we think about natural resources, energy use, pollution, fairness, and sustainability. As pressures mount from population growth, climate change, ecosystem degradation, and the increasing likelihood of sudden changes in Earth's behavior, a growing number of business leaders have discovered that a safe pathway into the future can also be a profitable one.

Walking down the snow-covered streets of Davos, Switzerland, during the annual World Economic Forum these days, one can't help but notice that every company wants to brag about its most innovative ideas. Whether it's the electric car manufacturer Tesla, VW E-Up, or Audi's E-tron, solar technologies, or new lightweight carbon materials, everybody's talking about innovation aimed at sustainability. Think about it: When a business really wants to show off today, it doesn't hang its logo on the equivalent of a gas-guzzling Hummer; it proudly puts forward its leanest, cleanest, and coolest new product.

Because it's the right thing to do.

1

OUR NEW PREDICAMENT

THE WORLD AS WE KNOW IT is a relatively new phenomenon. For most of Earth's 4.5 billion-year history, conditions on the planet have been far less hospitable than they are today. It has only been during the past 10,000 years, in fact, that factors necessary for human societies to develop have been reliably present. Before that, Earth was often a horror show.

When our ancestors, the earliest hominids, first appeared in Africa some 2.5 million years ago, for example, they faced a series of crises as Earth shifted back and forth between punishing ice ages and lush warm periods. Even when modern human beings began to walk the planet about 160,000 years ago, survival was still not a sure thing. The world's climate kept alternating between cold episodes of expanding ice sheets, water scarcity, low sea levels, and food shortages, and warm episodes of abundant water, high seas, and lush biomass resources. Although these swings weren't extraordinary from a geological perspective— global temperatures varied less than 5°C (9°F) one way or the other—their consequences were huge for human survival.

Back then, the relatively small human population, fluctuating between a few million and tens of millions, lived as hunters and gatherers. During periods of extreme climate shifts, when it was hard to find food and shelter, they were confined to pockets of productive savannahs in Africa. In a critical cold period about 75,000 years ago, as DNA analyses have revealed, the entire human population may have dwindled to as few as 15,000 fertile adults, confined to the high plateau in northern Ethiopia. This constituted a profound crisis for our species. We've never been as close—in fact, one cannot be closer—to extinction. To find new sources of food, groups of survivors set out along the coasts of the Red Sea, which at the time may have been as much as 100 m (328 ft) lower than it is today (because so much freshwater was tied up in the polar ice sheets). Walking first through the southern Arabian Peninsula—then, as now, an arid and inhospitable environ-

The loss of rainforest such as this in the Bandan River area of Sarawak, Malaysia, may affect moisture feedback from the forest, thereby changing regional rainfall patterns.

ment—and then moving along the coast toward India, these groups, among others, eventually spread to Australasia and Europe some 40,000 years later.

It was the scale and rapidity of climatic changes that locked humanity into a semi-nomadic lifestyle. Sometimes these changes were abrupt. As researchers have discovered from ice cores drilled deep into ice sheets on Greenland, some changes over the past 100,000 years took place in a matter of only decades. About 11,500 years ago, for example, temperatures in Greenland shot up by 5–10°C (9–18°F) over a period of barely 40 years.

But then, about 11,700 years ago, Earth's stormy climate tapered off, as we left the last ice age and entered a planetary state of natural harmony, the interglacial period we now call the Holocene. Humanity literally came in from the cold into a remarkably stable warm environment. Compared to what we faced during the Pleistocene 2.6 million years ago, humans now enjoyed relatively minor changes in climate. In fact, in both the Northern and Southern hemispheres, we quickly grew accustomed to an incredibly narrow range of climatic variation, with temperatures wobbling only about 1°C (1.8°F) up or down.

The impact was immediate. Almost as soon as we entered the Holocene, groups of hunters and gatherers in at least four different parts of the world independently invented agriculture more or less simultaneously. Clearly, the warm, wet, and predictable environment agreed with us. Within 1,000–2,000 years of this new regime, we saw a transformation of our way of life in many places from semi-nomadic hunting and gathering to sedentary farming, which proved to be the key to the development of modern societies. Agriculture allowed specialization, technology development, rules and norms, and dramatic growth in our capacity to provide food for populations.

It was at this point in history that we saw the rise of the earliest advanced human cultures: the Longshan Neolithic agrarian cultures of the Yellow River Valley in China; the ancient Egyptian irrigation societies along the Nile; the Mesopotamian irrigation societies along the Tigris and Euphrates rivers; the Greek and Roman empires; the Islamic civilizations in a large part of Africa and Central Asia; the agrarian societies of the Maya civilization in Central America. In their own ways, each of these agrarian cultures developed into advanced societies. Civilizations evolved through the Middle Ages, the evolution of the feudal merchant societies, eventually coupling with the rise of modern science during the late Renaissance, and ultimately the birth of nation states. We reached our first billion people by the year 1800, growing to three billion by the middle of the 20th century.

The onset of the Holocene, in short, was the planetary equivalent of establishing the ultimate shopping mall for humanity. Suddenly we had a reliable source of goods and services delivered from a stable equilibrium of forests, savannahs, coral reefs, grasslands, fish, mammals, bacteria, air quality, ice cover, temperatures, freshwater availability, and productive soils. The point of this story is as simple as it is dramatic: We still depend on the Holocene for our prosperity and wellbeing. It is the Garden of Eden for our civilizations. In fact it's the only state of the planet we know that can support modern societies and a world population of more than seven billion people.

That's why what we're doing right now ranks as the most disturbing event in the history of humankind: We're pushing our planet out of the Holocene into new and uncharted territory.

WELCOME TO THE ANTHROPOCENE

It didn't take long—only a half-century or so—for the rapid pace of industry and agriculture to threaten the world as we know it. Since the great acceleration of the human enterprise, kicking off in the mid-1950s, humanity's wide-ranging impacts—including climate change, chemical pollution, air pollution, land and water degradation, nutrient overload, and the massive loss of species and habitats—have put nearly all of Earth's major ecosystems under stress. In fact, we humans, *Anthropos* in ancient Greek, have become such a massive source of global change that we now constitute a geological-size force on the planet, one even more extensive in magnitude and pace than volcanic eruptions, plate tectonics, or erosion. With reckless abandon, we've introduced our own geological epoch, the "Anthropocene."

The trend started in the mid-18th century with the industrial revolution, when we learned how to exploit fossil fuels as a new, cheap, and effective energy source. This broke many constraints that had previously hampered social and economic development. Now we could clear land in an unprecedented way, changing a landscape almost instantly. We developed an industrial process, only possible with fossil-fuel energy systems, to convert nitrogen from the atmosphere into fertilizers, breaking a fundamental constraint on food production. We improved sanitation systems, which, along with major medical advances, yielded great benefits for human health and improved urban environments. Our populations grew rapidly as a result of higher life expectancy and wellbeing.

Manufacturing systems emerged that used fossil fuels to greatly increase production of goods. Unknown to us at the time, this rapid expansion of fossil-fuel usage was slowly raising CO_2 concentrations in the atmosphere. By the early 20th century, CO_2 concentrations had reached the highest limits since the Holocene epoch began. Even back then, in other words, we were saying goodbye to the world as we know it.

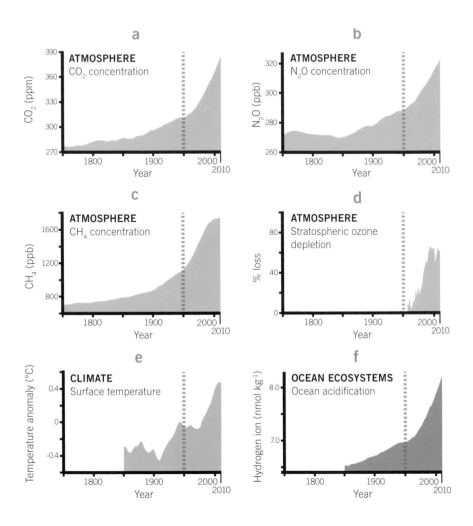

Figure 1.1 The Great Acceleration of Human Pressures on the Planet Starting in the mid-1950s, there has been a rapid increase in changes, on a global scale, to all the environmental processes that form the basis of our modern economy—all of it caused by human activity.

These curves reveal accelerations in the following: (a) the atmospheric concentration of CO_2, (b) the atmospheric concentration of N_2O due to agriculture and burning fossil fuels, (c) the atmospheric concentration of CH_4 due to the expansion of livestock-production systems, (d) the percentage of ozone-layer loss due to ozone-depleting chemicals used by humans, (e) the anomalies in average temperatures in the Northern Hemisphere, (f) the increase in ocean acidification,

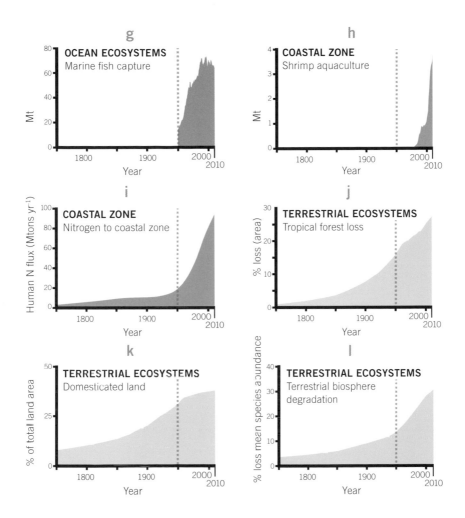

(g) the percentage of global fisheries either fully exploited, overfished, or collapsed, (h) annual shrimp production as a proxy for coastal zone alteration, (i) nitrogen pollution of coastal areas, (j) the loss of tropical rainforest and woodland in tropical Africa, Latin America, and South and Southeast Asia, (k) land converted to pasture and cropland, and (l) the estimated rate of extinction of species on Earth. (The abbreviations used above are as follows: ppm/ppb is parts per million/ billion, nmol kg is nanomole per kilogram, and Mt is million tons.)

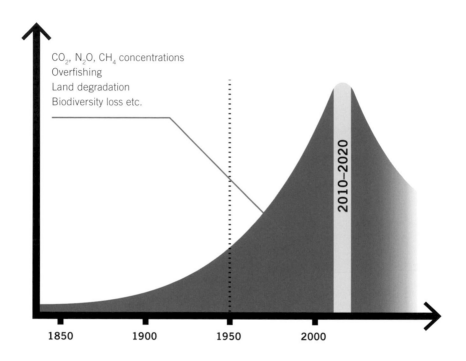

Figure 1.2 Exponential Impacts. The world embarked on the "Great Acceleration" of the human enterprise in the mid-1950s. At that time, the global population was about three billion people, with fewer than 0.4 billion having a lifestyle that significantly and negatively affected the global environment. At the same time industrial growth affected multiple environmental processes, leading to an exponential rise in negative pressures. Today we are at the top of that curve, maintaining the pressure on the planet. More and more science shows that we have reached a saturation point. We have hit the ceiling of what the planet can cope with, risking the onset of catastrophic changes. We urgently need to reverse the trend of detrimental global environmental change. This needs to happen in the current decade for most environmental processes in order to provide stable "Holocene-like" conditions in which we can secure prosperity in a still-growing world.

It wasn't until the mid-1950s, though, a period we now refer to as the "Great Acceleration," that the impacts of humankind's activities grew to dangerous levels. There were still relatively few people on the planet at that point—about three billion in 1955—and we were still working under the mistaken assumption that environmental issues were separate from social and economic ones. But soon, by almost every measure, our growing populations and unsustainable habits piled on more and more environmental pressures. No matter which parameter you chose to look at—whether CO_2 concentrations in the atmosphere from fossil fuels; nitrogen concentrations in the soil from agriculture and industry; methane concentrations in the air from livestock; ozone depletion over Antarctica; rising surface temperatures; floods and other extreme weather disasters; disappearing fish stocks; coastal disruption from fish farms; nitrogen pollution of coastal waters; loss of tropical rainforests; wild habitat converted to croplands; or increased rates of biodiversity loss—the trend was the same, with a sharply rising curve heading in the wrong direction (see Figure 1.1).

Today, having reached unprecedented levels, these pressures have generated a vise of global impacts we call the "Quadruple Squeeze" (see Figure 1.4). The first squeeze on our wellbeing comes from the quest for affluence on an increasingly crowded planet. Of the nine billion people predicted to inhabit Earth by 2050, almost all of the population growth is expected to take place in what are today poor communities in Asia, Latin America, and Africa. At the same time, as the latest assessments by the Organisation for Economic Co-operation and Development (OECD) show, the world economy is projected to almost triple by 2050. Most of this growth is expected to occur among the world's poorer nations, whose economies are projected to expand five-fold. We'll soon live in a world with not 1.5 billion, but rather four, five, or even six billion middle-class citizens. For the first time in modern history, it is possible to imagine a world in which absolute poverty has been eradicated.

This would be absolutely wonderful, and a grand marker of the right of all citizens to development and a decent life, but it entirely changes the prospects for human economic development. Many of the world's resources have already been used by industries and nations that have exploited them as quickly as possible to benefit the wealthiest 20 percent or so of the global population. Now the remaining 80 percent, who have claimed little ecological space so far, are rightly claiming their own share of Earth's resources. The problem is, this 80 percent aspires largely to the same unsustainable lifestyles as the wealthy 20 percent— the same social and economic paradigm that has caused our problems so far, born out of an inherited lack of environmental knowledge, responsibility, and concern.

A pair of putty-nosed guenon, tree-dwelling monkeys, are offered for sale as bushmeat by poachers along a roadside in Cameroon.

This is where world leaders have tended to put their heads in the sand. It's apparently too painful for them to admit that, just when we're finally getting a chance to go to scale with economic growth for a majority of the world's citizens, they're being forced to recognize that the old party is over. We're hitting the hard-wired biophysical ceiling of Earth's capacity to support continued unsustainable growth.

What we need now is a deep rethink, a total mind-shift about the way that our economies should develop within the life-support systems on Earth. If we're serious about social and economic development for everyone, it must be founded on principles that are not only safe but also include a fair and just sharing of Earth's remaining ecological space among all citizens today and in the future. This is a massive, and so far heavily underestimated, collision between meeting peoples' needs and desire for affluence on the one hand, and securing a future within planetary boundaries on the other.

The second squeeze on our prospects comes from climate change. Since 1960, global CO_2 emissions have jumped from about 4 billion tons of carbon a year to about 9 billion tons. This represents a very rapid growth, with the biggest amounts coming during the past 15 years, which is paradoxical, of course, since that was the only period in human history when governments have agreed to reduce emissions. Meanwhile, concentrations of CO_2 in the atmosphere have risen from 280 parts per million (ppm) in pre-industrial levels to 400 ppm in 2014—the widely recognized ceiling with regards to acceptable climate risk. This represents, for all greenhouse gases (GHG), a concentration of approximately 450 ppm (CO_2 equivalent), the highest in at least 800,000 years.

Because of the complexity of the climate system, it's impossible to predict the exact amount of warming this increase in greenhouse gas emissions will cause. But the common understanding among scientists is that we can expect an average global temperature bump of about 3°C (5.4°F) if CO_2 concentrations rise as high as 560 ppm—that is, a doubling compared to pre-industrial concentrations. Unfortunately, the latest fifth scientific assessment of the Intergovernmental Panel on Climate Change (IPCC AR5) shows that we are even exceeding this risk, and following a path toward 4°C (7.2°F) warming by 2100, based on the momentum of global increases in greenhouse gas emissions and the lack of progress by world leaders in climate negotiations. This path can't be called anything but disastrous for humanity.

The world has already started to feel the consequences of rising temperatures. In fact, the signs are all around us: rapid loss of summer sea ice in the Arctic Ocean; retreat of mountain glaciers around the world; accelerated melting of the Greenland and West Antarctic ice sheets; an increased rate of sea level rise;

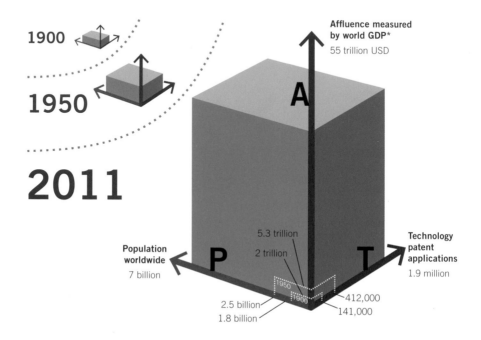

Figure 1.3 The Impact of Affluence. Three dimensions are often used as indicators of human social pressure on the planet: the number of people (P), affluence (A), and the level of technology (T). According to a commonly used formulation known as IPAT, human impact (I) is a result of the combined effects of P, A, and T. Using world GDP as a proxy for affluence and the number of technology patents as a proxy for technology, this figure (adapted from *National Geographic*, March 2011) illustrates the evolution of the human global squeeze. Until 1900, the human squeeze was confined within the smallest box. Between 1900 and 1950 the cumulative squeeze increased only incrementally. From 1950 onward the world "exploded," exerting a cumulative human squeeze corresponding to the entire volume of the largest box. But, it is not only size that matters. During the period before the "Great Acceleration" the human global squeeze was largely attributable to population growth—technology and economic growth mattered less. From the 1950s onward, the primary driver of the squeeze is affluence. This causes a major impact on the planet because we have used our wealth to buy "stuff," which in turn generates environmental impacts that we now need to stop and reverse.

*GDP, or Gross Domestic Product, is the total value of goods produced and services provided in a country in one year.

and an increase in bleaching and mortality in coral reefs. During the past few years, we've also seen an exceptional number of extreme weather events, many of which were likely strengthened by climate change. After 12 years of drought, Australia was struck in 2010 by the largest flood in 50 years. Dams overflowed and agricultural yields collapsed in a disaster that caused a significant reduction in Australian gross domestic product (GDP) and influenced world market prices for food. Unprecedented floods swamped Pakistan, India, and Afghanistan, while droughts caused social disruptions in parts of West and East Africa. In the USA, a record 14 weather and climate disasters caused damage estimated at 1 billion USD or more, from extreme heat waves, droughts, and floods to tornadoes, hurricanes, wildfires, and winter storms. Clearly, changes in rainfall patterns could represent a serious threat if region after region suffers shifts in the frequency, magnitude, and duration of droughts, wildfires, storms, floods, and disease outbreaks, that, in turn, affect food production, trade, economic growth, and, ultimately, social stability.

The third major squeeze on global progress comes from the extraordinary pace with which we are undermining Earth's biosphere—the marine, freshwater, and terrestrial ecosystems upon which all human societies depend. Never before have we eroded ecosystem functions and services as rapidly as we have during the past 50 years. Fish stocks have disappeared. Fish farms have disrupted coastal ecosystems. Coastal waters have been polluted with nitrogen and phosphorus. Tropical rainforests have been lost. Wild habitats have been converted to croplands, and we've seen dramatic reductions in biodiversity. In plain language, we've driven Earth to its weakest state since the advent of modern human societies, unwittingly limiting our options for the future.

The fourth "squeeze" on humanity's "room for maneuvering" comes from the recent recognition that sudden, unexpected change appears to be the rule rather than the exception in natural ecosystems. Although we've built our entire relationship with nature—our governance systems, our economic paradigm, and our regime for resource use—on the assumption that the environment functions like a well-stocked mall, with goods and services *gradually* and *linearly* becoming scarce as we exploit them, we now know that's not the case. If we overuse something in nature, it doesn't automatically restore itself like consumer products on store shelves appear to do. That's not how nature works.

During the past 30 years, research in resilience and complex systems has provided increasing evidence that the contrary seems to be true. Instead of continuous, slow incremental change, systems that over very long periods of time show signs of only gradual change can abruptly shift in pervasive and often irreversible ways. As Steve Carpenter, a researcher in aquatic ecosystems,

THE QUADRUPLE SQUEEZE

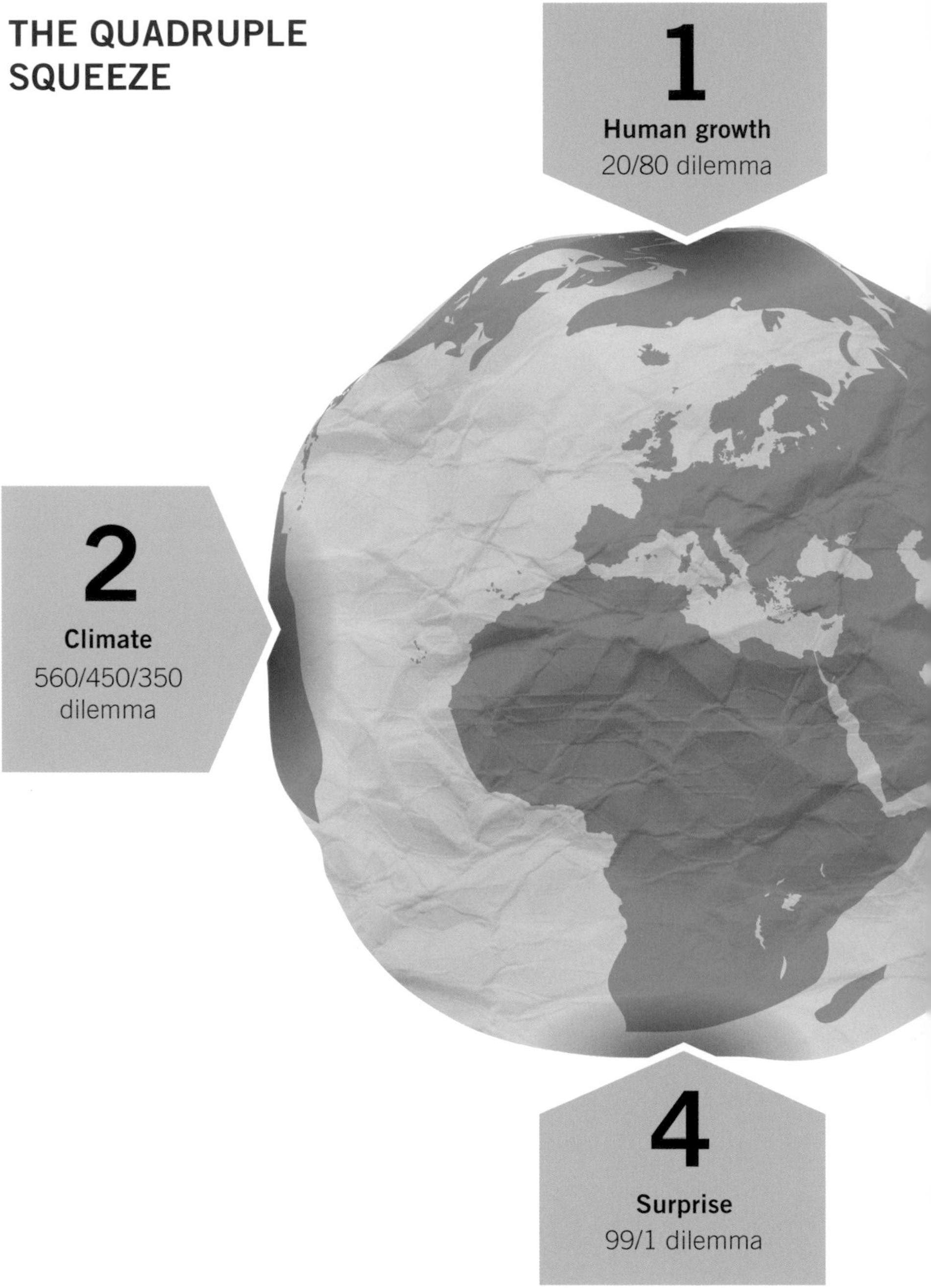

1
Human growth
20/80 dilemma

2
Climate
560/450/350
dilemma

4
Surprise
99/1 dilemma

3

Ecosystems
60% loss dilemma

Figure 1.4 The Quadruple Squeeze Humanity's ability to secure long-term sustainable development is under pressure from four planetary squeezes: (1) population growth and affluence, (2) climate change, (3) ecosystem degradation, and (4) the risk of sudden change or surprise when ecosystem thresholds are crossed, which reduces the operating space for human development.

The first squeeze derives from population pressure, where the dominating force is affluence. The bulk of environmental problems so far have been caused by about 20 percent of the world's population, the rich minority. The remaining 80 percent also have a right to pursue development.

The second squeeze is because we have almost reached 450 ppm of greenhouse gas concentration (CO_2 eq) at a rate that suggests we are rushing toward 560 ppm, a doubling since the industrial revolution, even though science tells us we must not exceed 400 ppm if we want to avoid large risks.

The third squeeze is that we are rapidly eroding the resilience of Earth, having already undermined 60 percent of key ecosystem services in support of human wellbeing.

The fourth squeeze comes from the shrinking safe space for human development. We need to recognize that abrupt change is commonplace in ecosystems and that the way to deal with this is to build in redundancy and buffering.

ppm (parts per million) is the measure of concentration of greenhouse gases in the atmosphere.

CO_2 eq (Carbon dioxide equivalent) is a standard unit for measuring carbon footprints. The idea is to express the impact of each different greenhouse gas in terms of the amount of CO_2 that would create the same amount of warming. Apart from CO_2 they include methane (CH_4), Nitrous oxide (N_2O), Ozone (O_3), and chlorofluorocarbons (CFCs).

has put it: "Ninety-nine percent of change in ecosystems appears to happen as a result of one percent of events affecting the system."

This alarming recognition—which will be further explored below—that nature often behaves in unpredictable ways, with the risk of major and often irreversible shifts from one state to another, fundamentally changes our relationship to the planet. To preserve the stability of biophysical systems, and thus Earth's ability to support human development, we must invest in building ecological resilience—nature's capacity to avoid undesired surprises. That requires investment in diversity and redundancy, in our ecosystems, biomes, and in our Earth system as a whole. We need ecological shock absorbers to buffer the stress of unexpected impacts.

These four squeezes, then, are the main pressures limiting our options for the future. It's not difficult to see how we got here. Population growth, climate change, ecosystem degradation, and abrupt change are the flipside of a coin that has given us unprecedented social benefits. The rising curves of negative environmental change have a sister curve in rising human wealth. The latter is a remarkable success story. Despite the fact that absolute poverty remains stubbornly high, affecting about a billion men, women, and children, more people are wealthier than ever before in modern history. We produce 1.5 times more staple foods today than we did 50 years ago, outpacing population growth. We also use nearly four times as much energy. The average life expectancy in most nations exceeds 65 years, and women on average give birth to close to two children, two critical indicators of great success in improved human wellbeing.

In terms of social benefits, one could argue, it has been worth it. Environmental degradation, one could say, is simply the price we must pay to create wealth in a world with seven billion people. Unfortunately, this argument no longer holds true, and probably never did. We now know that sustainability, not over-exploitation, is the real basis for social wellbeing.

EVERYTHING IS CONNECTED TO EVERYTHING ELSE
The world as we know it has become an increasingly complex, turbulent, and globalized place, not only socially and economically but also ecologically. What we do in one corner of the planet now affects, in real time, living conditions for fellow citizens in other parts of the planet. The way we go to work in Sweden influences rainfall patterns for a small-scale farmer in southern Africa, and

Preceding pages: Dubai City, in the United Arab Emirates, reflects the challenges of sustainable urban development—key to dealing with squeezes from population growth and affluence.

the way fishermen in Thailand manage mangroves affects weather patterns in England. The social–ecological wiring of the modern economy is now global.

Because of that, we can no longer act locally to further global development; we must also act globally to further local development. No matter how well environmental policies are applied in places such as the Galapagos, West Papua coral reefs, or the Arctic, their continued success now depends on the actions of other nations, regions, and economic sectors. Environmental protection can only succeed in combination with planetary stewardship.

Consider the unexpected impacts of forest losses in Borneo. During the past 20 years, the area under palm oil cultivation in Indonesia and Malaysia has roughly tripled, helping to accelerate—along with logging operations, soya bean cultivation, bauxite mining, and livestock operations—the destruction of the region's remaining rainforests. The loss of these ecosystems, rich in biodiversity, has not only eliminated wildlife habitats, it has also undermined local communities, which depend on the rainforest for small-scale agriculture, forest management, and fisheries. Most people living in the Borneo rainforest get their protein not from the forest itself, but from the rivers flowing through it. Palm oil plantations load up the rivers with sediment caused by soil erosion, gradually destroying the waterways with nutrient overload and pesticides. This degrades fish stocks, which undermines the local people's access to protein. If salaries aren't high enough for workers to afford legal food sources, what happens next is an increase in the illegal bushmeat trade and, in time, a severe threat to vulnerable and endangered species.

Until recently, such stories didn't mean too much to the rest of us. If industrial agriculture disrupted a distant community living off healthy local ecosystems, it was dismissed as a development failure—a local problem. But now this is changing. In an increasingly interdependent world, what happens in Borneo is affected by and affects societies everywhere. The notion that everyone lives in everyone else's backyard is the new reality.

This was dramatically demonstrated in 2008 when massive forest fires broke out in Borneo. The island's forest biome has always relied on the relatively dry conditions of a roughly four-year climate cycle to trigger re-growth of plants and trees. But deforestation has opened up landscapes even further. Combined with climate change, which has increased the likelihood of drought, deforestation has introduced fire to a region that had no prior fire regime. In fact, the forest fires that broke out in 2008 were so extensive they created an Asian "brown cloud" across Southeast Asia, contributing the equivalent of 30 percent of that year's global emissions of greenhouse gases. Indonesia, which used to capture carbon with its vast rainforests, was suddenly one of the largest greenhouse gas polluters

BALL-AND-CUP DIAGRAM OF RESILIENCE IN SOCIAL–ECOLOGICAL SYSTEMS

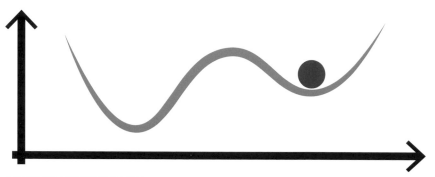

1 DESIRED RESILIENT STATE

A social–ecological system, such as a farming system, an urban region, or a coral reef, has an inbuilt capacity to take a hit—such as a drought or a financial crisis—and still "remain standing." The deeper the cup, which represents the resilience of the system, the larger is the chance that the system, represented by the ball, will remain in the desired state after being hit by a sudden shock. A resilient system has a better chance of remaining in a desired state. It also has the capacity to adapt to changing conditions.

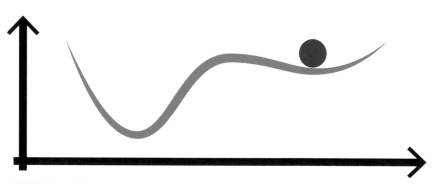

2 GRADUAL LOSS OF RESILIENCE

A system—such as a household, economy, forest, or polar ecosystem —gradually loses its ability to deal with shocks and stresses if it is not managed in a sustainable way. Examples of processes that erode resilience include indebtedness, excessive financial risks, loss of trust, or overdependence on a social safety net in human society; or loss of biodiversity, overuse of natural resources, climate change, or nutrient loading.

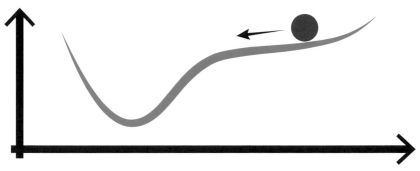

3 STATE SHIFT

A state shift occurs when a trigger or shock—such as a disease, drought, or flood—hits a system with low resilience. Feedbacks that hold the system in its desired state are replaced by opposite feedbacks that irreversibly draw the system to a new state. In a rainforest, for example, self-generating rainfall can be replaced by self-drying feedbacks characteristic of a savannah.

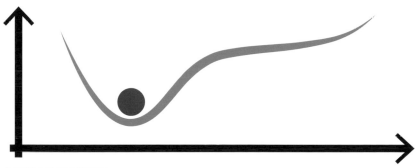

4 NEW STABLE STATE

The system gets locked into a new stable state if a reinforcing, positive feedback kicks in. Positive feedbacks accelerating climate change, for example, might include the release of methane from thawing permafrost, or melting ice lowering the planet's reflectivity of incoming solar radiation.

Figure 1.5 Tipping Points in Ecosystems. There are many examples of state shifts in ecosystems, such as when a coral reef ecosystem flips from one with hard coral to one with soft coral dominated by algae. Those coral reefs still rich in biodiversity today are subject to multiple pressures from overfishing, climate change, and nutrient loading from agricultural runoff. This weakens the coral reef, and a bleaching event, disease, or hurricane can be enough to push the ecosystem across a threshold and flip it into a soft reef system dominated by algae. This actually happened in the 1998 El Niño event which warmed the Indian and Pacific Oceans above a critical threshold, causing a major bleaching event. Up to 90 percent of coral reef systems collapsed in large parts of the Caribbean Sea and Indian Ocean. Many tipped over into a new state of soft coral and seaweed-dominated systems, when a new feedback kicked in, stimulating rapid dominance of seaweed. In some parts, a rich diversity of fish, particularly grazers, enabled coral reefs to bounce back and remain in the desired hard coral state by grazing down the onslaught of seaweed and thereby enabling the regeneration of hard coral.

Examples of Regime-Shifts in Ecosystems

Transgressing planetary boundary	Environmental change	Feedback causing tipping point	Ecosystem service and human wellbeing impacts
Climate change	Exceeding threshold of global Arctic warming causing accelerated melting summer ice.	Melting causes a change from white ice cover that reflects solar radiation to a darker water surface that leads to increased uptake of heat from incoming solar radiation, causing a self-accelerated ice melt.	Impacts climate regulation at a global scale, affecting agriculture, disease regulation, and cultural identities and lifestyles in potentially distant locations. Also expected to impact Arctic fisheries with consequences for industry and human livelihoods.
Nitrogen overload **Phosphorus overload**	Modern agricultural practices and poor treatment of urban waste causing overload of nutrients in freshwater systems.	Lakes, wetlands, rivers, and groundwater systems buffer nutrient overload up to a point where a threshold is crossed, triggering nutrient-thriving plankton to explode, which reinforces algae blooms and anoxic conditions.	Reduces water quality, therefore impacting fisheries, drinking water, and recreational opportunities, and leading to direct impacts on the economy, human livelihoods, and health.
Land use **Freshwater use**	Climate change, land degradation, and overuse of water causes a drying out of tropical savannah landscapes.	When warming and dessication passes a threshold, the canopy cover of trees, bushes, and plants reduces drastically (due to water scarcity), causing an abrupt loss in moisture feeding back into the atmosphere, in turn reducing rainfall. This can cause a savannah to enter a self-drying vicious cycle and turn into a desertified steppe.	Impacts range-fed livestock production, crop cultivation, soil erosion, and regional climate regulation, with direct impacts on food security, human livelihoods, and the potential for social conflict.
Ocean warming **Ocean acidification** **Overfishing** **Nutrient overload**	Oceans take up heat as the atmosphere warms, and become increasingly acidic as they absorb more and more CO_2. Together with nutrient overload, coral reefs lose resilience to deal with shocks, such as El Niño events.	Collapse of hard coral systems results from gradual heating of the oceans and extreme warming events (such as El Niño), combined with nutrient overload and biodiversity loss (for example, overfishing), and leads to the establishment of soft coral systems or algae-dominated reefs that thrive on nutrients and inhibit hard coral recruitment.	Impacts fisheries, biodiversity, and coastline protection from wave erosion, with substantial consequences for local livelihoods, and nutrition, the tourism sector, and coastal infrastructure.

Before
After

51

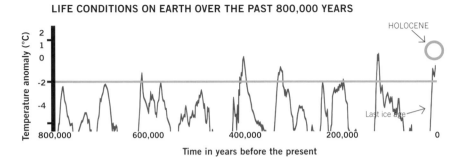

LIFE CONDITIONS ON EARTH OVER THE PAST 800,000 YEARS

Time in years before the present

Figure 1.6 The Last Time Earth Was 2°C Warmer Than Today. Ice core data from Antarctica shows the planet's swings in and out of ice ages over the past 800,000 years. The last time Earth was in a warm interglacial period was in the Eemian epoch about 120,000 years ago. For several thousand years, mean global temperatures were approximately 2°C (3.8°F) warmer than average pre-industrial temperatures. Sea levels then stabilized at 4–8 m (13–26 ft) higher than sea level today. This shows how sensitive Earth is to warming.

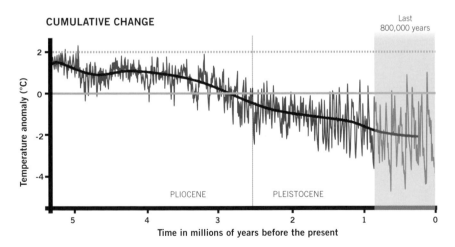

CUMULATIVE CHANGE

Time in millions of years before the present

Figure 1.7 Five Million Years of Large Variability. During the past two million years, Earth's climate has become colder and also more variable. In all this time, the average global temperature has never risen higher than 2°C (3.6°F) above the pre-industrial average—the level that world leaders have agreed that we should not now exceed, even though we know that major environmental changes will occur well before 2°C of warming.

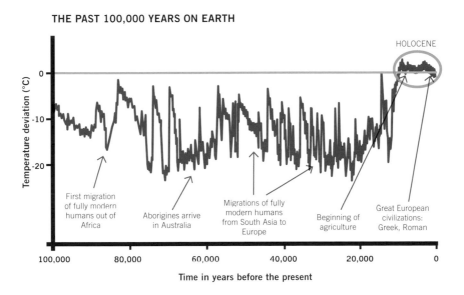

THE PAST 100,000 YEARS ON EARTH

HOLOCENE

Temperature deviation (°C)

0

-10

-20

First migration
of fully modern
humans out of
Africa

Aborigines arrive
in Australia

Migrations of fully
modern humans
from South Asia to
Europe

Beginning of
agriculture

Great European
civilizations:
Greek, Roman

100,000   80,000   60,000   40,000   20,000   0

Time in years before the present

Figure 1.8 Humanity's Period of Grace. Temperature in the Northern Hemisphere, as revealed by ice core data from Greenland, has been extremely variable during the past 100,000 years. Humans were fully modern during this period, but comprised only a few million hunters and gatherers. After this extremely bumpy environmental roller-coaster ride, we entered the extraordinarily calm and predictable Holocene epoch. During the 10,000 years that this era has lasted, humanity has enjoyed a global average temperature that has varied up or down by only about 1°C (1.8°F). Everything in nature that generates wealth and human wellbeing settled in during the Holocene, forming the basis for the emergence of modern civilizations. This figure also illustrates how much greater temperature swings have been in the polar regions compared to the planet as a whole. While the world has warmed approximately 0.8°C over the past 50 years, the Arctic has warmed at least twice as much.

in the world. The "brown cloud" also reduced the so-called optical depth of incoming solar radiation (one of the nine planetary boundaries), where the thick smog of air pollutants dims incoming solar radiation and functions as a mirror, reflecting back incoming sunlight to space. This cools the planet, creating the paradox of one environmental problem (aerosols from forest burning and combustion) camouflaging another (emission of greenhouse gases). But it also affects regional rainfall patterns, such as the Southeast Asian monsoon, which in turn affects the economies of Singapore, India, Hong Kong, and beyond.

SHOCK ABSORBERS FOR THE PLANET

Perhaps the most important thing we've learned about the biophysical behavior of the Earth system is that it regulates itself in a two-step approach. In the first step, when its resilience is high, it applies biological, physical, and chemical processes to resist change, applying so-called negative (dampening) feedbacks to persist in its original state—whether that state is a cold glacial equilibrium or a warm interglacial equilibrium. Then in the second step, as its resilience is lost and conditions reach a tipping point, a regime-shift occurs, sending the world on a journey toward accelerating heat or intense cold, with no turning back.

So far, in response to the massive disruptions caused by human activities, the planet has demonstrated an impressive capacity to maintain its balance, using every trick in its bag to stay in its current state. It has reduced the impacts of greenhouse gases, deforestation, and land degradation by absorbing substances, adapting ecosystems, and tweaking and turning food chains. Of the 9 billion tons of carbon we dump into the atmosphere every year, for example, Earth now absorbs about half in the oceans and land. This is undoubtedly nature's largest free ecosystem service to the world economy and humankind.

Unfortunately, there is growing scientific evidence that we may have reached a saturation point in terms of our pressures on Earth. And the scariest part of this situation is that the planet's response to further human disruption is not likely to be direct and predictable. The biggest global changes won't come from the triggers themselves, but rather from the positive feedback processes they unleash.

We've already seen the first signs of positive (reinforcing) feedbacks at a large scale. Our forests and other terrestrial ecosystems may be sequestering carbon from greenhouse gas emissions at a slower rate than before. Oceans are rapidly acidifying as they absorb CO_2 from the atmosphere, affecting marine life. Deforestation, loss of biodiversity, and overuse of freshwater and land resources have all remodeled the planet in a way that reduces its capacity to withstand disturbances such as climate change. This means that everything truly is linked to everything. We won't stand a chance of avoiding dangerous climate change beyond a 2°C increase in warming unless we both reduce emissions of greenhouse gases and manage the world's forests, grasslands, and oceans in a sustainable way.

In a hyper-connected world, crises propagated by natural feedbacks can spread rapidly—not only through ecosystems but also through political systems. In the new global architecture of interacting sectors—financial, political, agricultural, security, energy—a disturbance in one system or geographical region will propagate across boundaries, and ultimately grow to become a major global crisis. Again, surprise is the common denominator.

We thus need to be very careful before exhausting nature. The living biosphere, if handled with care, is an extremely effective—in fact our best—source of planetary-scale resilience. It provides us with an effective insurance policy against shocks caused by global environmental change, whether from natural or human origins. As an integral part of the Earth system, we could learn from nature's example and model landscapes of our own that work as sources of resilience. But so far, as our predominant avenue of development, we've chosen another path. We've built systems, from agriculture to urban areas, that have reduced our resilience, giving us large benefits in the short term, while creating vulnerabilities in the long term. This hasn't been a very intelligent strategy.

Although it might work for a while to support our wellbeing, it cannot be sustained indefinitely. A new strategy is urgently needed to slow, stop or even reverse the pace of global environmental change, while, at the same time, ensuring prosperity for a rapidly growing population. Accomplishing this while also addressing the needs of more than half of the world's population, for whom poverty remains a reality, will be challenging. But as we see it, the only way to succeed is to recognize the virtue of the Holocene—and to do everything we can to preserve this stable and beautiful state of the planet, which is the one state we know that can support the modern world as we know it.

Following pages: Fishermen crowd a small boat off Tanzania, facing dwindling fish resources in part due to illegal operations by long-distance fishing boats.

2

PLANETARY BOUNDARIES

IF YOU WERE TO FIND yourself driving down a winding road along a steep cliff on a dark night, you'd want clearly marked guardrails to prevent you from getting too close to the edge. That's the basic idea behind planetary boundaries. The human enterprise, as we've seen, is racing into the future at breakneck speed with hazards at every turn—hazards as dangerous and abrupt as plunging off a cliff. To avoid such disastrous outcomes for humanity we need to define planetary boundaries to act as guardrails to keep us from accidentally going over the edge. These boundaries won't hinder growth or development, just as guardrails along a meandering road don't slow down the progress of drivers. They're there to prevent a catastrophe.

We didn't need planetary boundaries in the 1980s, when we still lived in a small world on a big planet. We thought we could exploit minerals, living species, freshwater, land, oil, coal, and natural gas, without ever asking—not seriously, at least—whether infinite growth was possible on a finite planet. Now, all that has changed. Today we need a new framework for development that respects the true functioning of Earth's climatic, geophysical, atmospheric, and ecological processes, a development paradigm in which human prosperity and economic growth occur within the safe operating space of a stable and resilient planet.

The starting point for this quest—*to define a safe operating space for humanity on a stable planet*—is to identify which of Earth's processes are most important to maintaining the stability of the planet as we know it. As we saw in Chapter 1, there's ample scientific evidence that the Holocene, the warm and stable interglacial period humanity has enjoyed for the past 10,000 years is the only biophysical state that can support the modern world as we know it. What must we do to maintain Holocene-like conditions on Earth, now that we, in the Anthropocene, have become a global force of change?

About 40 percent of all the land on Earth has been converted to agriculture, such as rice cultivation in Vietnam's Mekong Delta.

We know the Holocene well. We know how the ocean, land, and atmosphere interact, and how the climate system interacts with the biosphere. We understand the hydrological cycle, which regulates the global amount of biomass, together with land, nutrients, and energy from the sun. We understand the global cycles of carbon, nitrogen, and phosphorus. We know that to maintain our planet in its current state of balance, we depend on permanent ice sheets in the North and South poles, which reflect back into space a large proportion of heat from the sun. We know that living ecosystems determine the flux of oxygen, water, carbon, methane, and other elements in the Earth system through landscapes and oceans. By using the Holocene as a reference point for our future on Earth, in other words, we can scientifically quantify the boundaries we need to avoid pushing our planet out of its, for us, desired state.

In pursuit of this finding, our interdisciplinary group of scientists, including Hans Joachim Schellnhuber of the Potsdam Institute for Climate Impact Research, Will Steffen of the Australian National University, Katherine Richardson of the University of Copenhagen, Jonathan Foley of the University of Minnesota, Nobel Laureate Paul Crutzen of the Max Planck Institute for Chemistry, and many others, evaluated the forces that regulate the way Earth functions. We looked at one process after another, exploring interactions between them, and seeking to characterize the conditions needed to remain in a stable Holocene-like state. For each system, we made the best possible attempt, based on the latest scientific evidence at hand, to quantify the biophysical limits outside which the system might flip into a different, and for us undesired, state. These were the planetary boundaries.

This quantification had never been done before. Only during the past 10–15 years have scientists been able to explain the complex dynamics governing the way Earth operates. The first comprehensive overview of observations showing the exponential rise of human pressures on the planet during the past 50 years was published only eight years ago. But the groundwork for the planetary boundaries concept rests on more than 30 years of empirical research showing that ecosystems, from local lakes to forest biomes and large ice sheets, can abruptly cross tipping points and irreversibly shift from one stable state to another, unless they're resilient enough to resist change. Whether we call them tipping elements, tipping points, regime-shifts, or thresholds is less important than the mountain of evidence showing that Earth's resilience is what matters.

Sure, there were warnings in the past, from Rachel Carson's *Silent Spring* in 1962 to the "Limits to Growth" analysis a decade later by Donella and Dennis Meadows, their colleagues, and the Club of Rome think tank. But these voices were drowned out by conventional economists, policymakers, and business

The 2014 Update on Planetary Boundaries

Earth System Process	Control Variables	Planetary Boundary (zone of uncertainty)	Current Value of Control Variables
Climate change	Atmospheric CO_2 concentration, ppm	350 ppm CO_2 (350–450 ppm)	396.5 ppm CO_2
	Energy imbalance at top-of-atmosphere, W/m^2	Energy imbalance: $+1.0$ W m^{-2} ($+1.0$–1.5 W m^{-2})	2.3 W m^{-2} (1.1–3.3 W m^{-2})
Biosphere integrity	Genetic diversity: Extinction rate	Genetic: < 10 E/MSY (10–100 E/MSY) but with an aspirational goal of ca. 1 E/MSY (the background rate of extinction loss). E/MSY = number of extinctions each year per million species	100–1,000 E/MSY
	Functional diversity: Biodiversity Intactness Index (BII) Note: These are interim control variables until more appropriate ones are developed	Functional: Maintain BII at 90% (90–30%) or above, assessed geographically by biomes/large regional areas (for example, southern Africa), major marine ecosystems (for example, coral reefs) or by large functional groups	84.4%, applied to southern Africa only
Novel entities	No control variable currently defined	No boundary currently identified, but see boundary for stratospheric ozone for an example of a boundary related to a novel entity (CFCs)	
Stratospheric ozone depletion	Stratospheric O_3 concentration, DU	<5% reduction from pre-industrial level of 290 DU (5%–10%), assessed by latitude	Only transgressed over Antarctica in Austral spring (· 200 DU)
Ocean acidification	Carbonate ion concentration, average global surface ocean saturation state with respect to aragonite (Ωarag)	≥80% of the pre-industrial aragonite saturation state of mean surface ocean, including natural diel and seasonal variability (≥80%– ≥70%)	~84% of the pre-industrial aragonite saturation state
Biogeochemical flows: (P and N cycles)	P cycle — Global: P flow from freshwater systems into the ocean	Global: 11 Mt P yr^1 (11–100 Mt P yr-1)	22 Mt P yr^1
	P cycle — Regional: P flow from fertilizers to erodible soils	Regional: 3.72 Mt yr^1 mined and applied to erodible (agricultural) soils (3.72–4.84 Mt P yr^1). Boundary is a global average but regional distribution is critical for impacts.	~14 Mt P yr^1
	N cycle — Global: Industrial and intentional biological fixation of N	44.0 Mt N yr^1 (44.0–62.0 Mt N yr^1). Boundary acts as a global "valve" limiting introduction of new reactive N to Earth system, but regional distribution of fertilizer N is critical for impacts.	~150 Mt N yr^1
Land-system change	Global: area of forested land as % of original forest cover	Global: 75% (75–54%) Values are a weighted average of the three individual biome boundaries and their uncertainty zones	62%
	Biome: area of forested land as % of potential forest	Biome: Tropical: 85% (85–60%), Temperate: 50% (50–30%), Boreal: 85% (85–60%)	
Freshwater use	Global: Maximum amount of consumptive blue water use (km^3yr^1)	Global: 4,000 km^3 yr^1 (4,000–6,000 km^3 yr^1)	~2,600 km^3 yr^1
	Basin: Blue water withdrawal as % of mean monthly runoff	Basin: Maximum monthly withdrawal as a percentage of mean monthly runoff. For low-flow months: 25% (25–55%); for intermediate-flow months: 30% (30–60%); for high-flow months: 55% (55–85%)	
Atmospheric aerosol loading	Global: Aerosol Optical Depth (AOD), but much regional variation		
	Regional: AOD as a seasonal average over a region. South Asian Monsoon used as a case study	Regional: (South Asian Monsoon as a case study): anthropogenic total (absorbing and scattering) AOD over Indian subcontinent of 0.25 (0.25–0.50); absorbing (warming) AOD less than 10% of total AOD	0.30 AOD, over South Asia region

leaders on the grounds that there was little or no evidence that humanity could push systems on Earth too far, or that we could exhaust the biosphere. Now, we know better. We have proof that human impacts on natural systems are hurting both our economies and our social welfare. We have evidence that critical thresholds are "hard-wired" into Earth's environment and that we'd be wise to avoid triggering them.

The biophysical components of the Earth system—what we call nature for short—are full of surprises, we've discovered, and if we push systems too far, they can break and get stuck in an undesired state for humanity. If such tipping points occur in too many systems in too many places—such as the irreversible melting of ice sheets or the release of methane in steppe regions from thawing permafrost—then the combined effect could lead to the crossing of a planetary tipping point that takes us away from the Holocene.

We now know that the causes of planetary tipping points in the distant past were not limited to factors such as changes in Earth's position relative to the sun or collisions with asteroids. They were also caused by interactions between external factors and responses from Earth itself. This is absolutely critical to recognize and something that science now understands with increasing precision. Earth is a complex and self-regulating system, in which everything is connected to everything else. This means, in very simple terms, that when nature is in good shape, Earth's resilience is high. When the climate is stable, rains are adequate, soils and the air are intact, then biodiversity is rich and ecosystems thrive. Being resilient enables Earth to apply its biophysical processes to dampen external impacts. It's like a boxer getting punched during a bout. In the first few rounds, when resilience is high, the boxer can absorb even strong punches without falling. But by the tenth round, as resilience is lost, the risk becomes much higher that the boxer, with the next punch, will cross a tipping point and get knocked out.

It's the same with Earth. Oceans, land, water, and biodiversity, through flows and stocks of energy, nutrients, carbon, and other elements, can dampen or reinforce the impacts of external punches. In the geologic past, these "punches" came from outer space. Today they come from us humans. Fundamentally, it is the same thing. Our emissions of greenhouse gases prompt a planetary-scale energy imbalance. The grand question is how does Earth respond?

So far, Earth has applied its biophysical resilience to absorb disturbances, hiding away 90 percent of the heat from greenhouse gas emissions in the ocean, and soaking up more than 50 percent of our CO_2 emissions in natural ecosystems. We call these negative feedbacks, since they reduce the impacts of disturbances. But as Earth's resilience is gradually lost, critical systems are likely to ultimately cross thresholds when feedbacks change direction from negative (dampening) ones to positive (reinforcing) ones. Quite abruptly, Earth changes from friend to foe, from sweeping abuses under the carpet to becoming a reinforcing engine of change.

That's why we need to respect planetary boundaries.

Having said that, defining safe boundaries is difficult because of the broad ranges of uncertainty involved. These are largely due to the difficulty of predicting how boundaries interact and respond to changes (such as what happens to the jet stream when the atmosphere traps more heat). With respect to human-induced climate change, for example, a vast majority of scientists now agree that we're likely to cross a critical threshold in Earth's climate somewhere between 350 to 450 ppm of CO_2, beyond which we'll very likely push the planet outside of our desired state.

The difference between 350 and 450 ppm encompasses a wide range of uncertainty. In the pre-industrial age (before 1750), the global atmospheric concentration of CO_2 was 280 ppm, a level that hadn't been exceeded for at least 100,000 years. Going beyond the range of uncertainty (higher than 450 ppm for CO_2) will almost certainly push us outside Holocene stability. If that happens, we're taking a big risk of triggering the irreversible melting of the Greenland and Arctic ice sheets, the West Antarctic ice sheet, the collapse of many tropical marine systems, the destabilizing of monsoon weather patterns, and so on. Having passed the 350 ppm mark around 1990, we've already seen some of the predicted consequences, such as the sudden reduction of ice cover in the Arctic Ocean, and the alarming findings in mid-2014 that several glaciers in West Antarctica have likely already crossed a threshold of irreversible melting, potentially adding another meter (3.3 ft) of sea level rise to the meter already expected by the end of this century.

And yet, despite such extremely troubling signals, science still cannot tell us precisely where the climate threshold lies. One reason is that our scientific understanding of how the Earth system operates is still incomplete, despite major advancements. Another reason is the fact that key processes interact with one another, which can affect their respective boundaries, such as the way that deforestation affects the global carbon sink, which in turn affects the safe level of CO_2 concentration in the atmosphere. If the question is how far we can push these processes before we actually cross a dangerous threshold, science will never be able to give us exact answers. That makes it extremely difficult to say at exactly what point humanity crosses a dangerous line.

This is not surprising. In a complex world we rarely if ever know things for certain. Instead we act on the best available knowledge to make a sober judgment of the risks. This is what planetary boundaries are all about, presenting the best possible scientific estimate of how we can avoid unacceptable risks of triggering catastrophic shifts in the Earth system.

Our approach to uncertainty is to apply principles of precaution. As the editors of *The Economist* pointed out not long ago with razor-sharp British understatement, "when changes occur faster than theory stipulates it should, a certain degree of nervousness is a reasonable response." In the same way, our group of scientists decided to place the safe boundaries for key processes at the lower end of scientific uncertainty—farther away from the cliff. In the case of climate change, that meant putting the boundary for CO_2 concentration at 350 ppm. Placing the guardrail in such a cautious position, of course, involved a normative judgment about the level of risk the global community is willing to take—how close to the edge we're all willing to drive our global economy. It also involved a judgment about how resilient we think societies might be to the consequences of crossing a tipping point, such as whether the world's coastal cities will be able to cope with several meters of rising sea levels if the climate change boundary is ignored. As with any good management or governance system, of course, we assumed the planetary boundary framework needs to be adaptive, continuously updated, and fine-tuned as new information is gathered, as things change.

Intuitively, such an approach feels obvious to most people. We're accustomed to operating within the confines of safe boundaries in our own lives. But so far we've remained remarkably negligent about doing so for the planet. We've never approached human development from the perspective of striving to remain in Holocene-like conditions. Rather than focusing on resource use, driving forces, or other direct aspects of human activity, we've never taken seriously the risk of abrupt or irreversible changes in the Earth system,

With respect to chemical or air pollution, for example, policymakers have customarily set targets at the local-to-regional scale, where they have attempted to define safe "critical loads," or "minimum standards," or "tolerable limits." Communities and other localities have typically done the same thing with acceptable loads of heavy metals such as lead or cadmium in drinking water. Efforts like these have been aimed at determining how much we humans can cope with, a compromise based on an assessment of the costs and benefits to human health, or they've tried to establish a load limit to protect the environment, to preserve a sliver of land or water as an oasis.

The planetary boundaries approach does away with such compromises by focusing directly on biophysical processes—especially the role of thresholds in triggering abrupt and potentially catastrophic changes—rather than on resource use and demand, driving forces, or other aspects of human activity. Moreover, it avoids the trickiest issue of all—estimating the degree of human

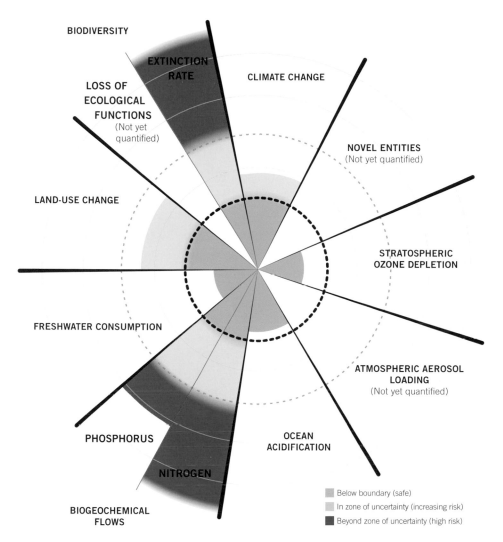

BIODIVERSITY

EXTINCTION RATE

CLIMATE CHANGE

LOSS OF ECOLOGICAL FUNCTIONS
(Not yet quantified)

NOVEL ENTITIES
(Not yet quantified)

LAND-USE CHANGE

STRATOSPHERIC OZONE DEPLETION

FRESHWATER CONSUMPTION

ATMOSPHERIC AEROSOL LOADING
(Not yet quantified)

PHOSPHORUS

OCEAN ACIDIFICATION

NITROGEN

Below boundary (safe)

In zone of uncertainty (increasing risk)

Beyond zone of uncertainty (high risk)

BIOGEOCHEMICAL FLOWS

Figure 2.1 Planetary Boundaries, the 2014 Update. The nine boundaries proposed in the original 2009 study have been scientifically reconfirmed, with updated boundary estimates and a few important adjustments of phosphorus and land-use change. The analysis, drawing on decades of advancements in Earth system science and resilience research, identifies the biophysical "safe operating space" on Earth to maintain a stable and resilient "Holocene-like" planet, within which humanity stands a high chance of thriving in the future. The green wedges inside the dotted circle depict the "safe operating space." When a boundary is transgressed beyond the safe operating space, we enter a danger zone of uncertainty, shown in yellow. Beyond the uncertainty range for science, we enter a high-risk zone for irreversible change, shown in orange. The wedges show the current position for each boundary.

ingenuity. No assumptions are made about our capacity for innovation, for taking technological leaps that could transform our world, either decisively toward sustainability or away from the Holocene. Instead, it simply defines a safe operating space for humanity, marking the planetary playing field within which humanity can innovate, pursue social and economic aspirations, experiment with different technologies, and apply different governance and political systems. It leaves plenty of room for a myriad of options, as long as we live on a planet in good shape.

Science has already been moving in this direction. With respect to climate change, for example, world leaders in 2009 agreed to limit global warming to 2°C above pre-industrial levels. This was a higher limit than our planetary boundary estimate of 1.5°C, but it was still a global climate boundary. Great research from colleagues at the Potsdam Institute for Climate Impact Research, the International Institute for Applied Systems Analysis, the Netherlands Environmental Assessment Agency, and the Environmental Change Institute at Oxford have subsequently translated this climate boundary to a global carbon budget. Their research has shown that we can emit no more than 1,000 billion tons of additional CO_2 into the atmosphere to maintain a 66 percent chance of staying below 2°C. That leaves us with no more than 25–30 years to transition away from a fossil-fuel-based world economy. This is applying planetary boundary thinking to one of the nine boundaries.

The game has thus changed.

THE NINE PLANETARY BOUNDARIES

Following a comprehensive review, engaging global change scientists from different fields, our team identified a number of planetary boundaries. Given the complexities and interactions that exist between all living and non-living components, from bacteria to bedrock, you might have expected we'd end up with 50 or more processes. Instead our critical assessment identified the following nine:

Climate change	*Freshwater consumption*
Stratospheric ozone depletion	*Land-use change*
Rate of biodiversity loss	*Nitrogen and phosphorus pollution*
Chemical pollution	*Air pollution or aerosol loading*
Ocean acidification	

A lake reflects the autumn finery of a forest in Dalarna, Sweden. Boreal forests are sensitive to temperature variations as shown by long-term forest carbon studies.

We published our findings in 2009 as a challenge to the scientific community. Had we missed something? Had we included any processes that clearly were not planetary boundaries? We put our findings out for scrutiny by the international science community, and, exactly as we hoped, it triggered a wide engagement by scientists around the world, as well as among policymakers, business leaders, and individuals from civil society. Five years down the line, following many scientific articles assessing and improving our original work, we feel more confident about our choices than ever. In 2014 we prepared a scientific update on the planetary boundaries, in which we concluded that the nine boundary processes we identified are the correct ones.

Identifying the nine boundary processes was one thing. Quantifying safe boundary levels was another altogether. In our 2009 assessment, we proposed quantifications for seven of the nine boundaries. For a few of these, the science was so robust and well advanced that boundaries were already well established. This was the case with climate change, where we set the boundary at 350 ppm CO_2 or 1 watt per square meter of additional human-caused climate forcing, and for stratospheric ozone depletion, where we proposed to keep the protective ozone layer intact by not allowing it to thin more than 5 percent compared to its pre-industrial thickness. For land-use change, global freshwater use, and ocean acidification, we felt relatively confident with otur proposed boundary levels, while recognizing that there was a very wide range of uncertainty, and therefore room for improved quantifications as science advances, and even for identification of better control variables. For the rate of biodiversity loss and human interference with the global nitrogen and phosphorus cycles, we were only able to provide first best guesses based on the evidence at hand. This has in turn resulted in important advancements in the science, which has resulted in greatly improved quantifications of these boundaries. Finally, we were never able to provide quantifications for the aerosol loading and chemical pollution boundaries. The reason was primarily because they depend on so many interacting factors, a myriad of different chemical compounds and pollutants.

Today, all the boundary quantifications have been updated based on the latest science, incorporating the key improvements proposed and published in peer-reviewed literature during the past five years by independent scientists around the world. We've made a first attempt at providing a good indicator for the aerosol loading boundary by quantifying a regional level of minimum optical depth of solar radiation for Southeast Asia. We're still struggling with the chemical pollution boundary, which we've renamed "novel entities" to indicate that these are truly novel compounds in the Earth system, created entirely by us humans.

One way to think about the boundaries is to put them in three groups, according to how they operate. The first group includes processes with sharply defined global thresholds, such as the risk of melting the Greenland and Antarctic ice sheets—processes capable of sharp shifts from one state to another, with direct implications for the entire planet. We called these "The Big Three." They are climate change, stratospheric ozone depletion, and ocean acidification. Thresholds for these processes are hard-wired into the Earth system and cannot be shifted by human actions (at a certain temperature the great ice sheets melt, and once this occurs, the planet will move away from its current Holocene equilibrium).

The second group includes boundaries based on slow planetary variables that contribute to the underlying resilience of the Earth system. We called them the four "slow boundaries." They are land-use change, freshwater use, biodiversity loss, and interference with the global nitrogen and phosphorus cycles. Unlike the first group of boundaries, which impact the Earth system from the "top down," the second group works from the "bottom up." Rather than being associated with change at the global scale, they're associated with local-to-regional scale thresholds. We see them as processes that operate "under the hood" of the planetary machinery to buffer the Earth system from harmful impacts and to strengthen its resilience.

Even though the boundary processes in this second group don't appear to have their own thresholds at the large regional or global scale, that doesn't mean there aren't thresholds associated with them. On the contrary, there's ample evidence that gradual changes in key variables such as biodiversity, harvesting of biomass, soil quality, freshwater flows, or nutrient cycles can trigger abrupt changes when critical thresholds are crossed in ecosystems such as lakes, forests, or coral reefs.

The question is: When do such tipping points become a global concern? When do they pose a threat to Earth's ability to stay in a Holocene-like state? If overfishing and nutrient loading cause a few lakes in one region to collapse, that might not have any implications for a region or Earth as a whole. But if such tipping points occur simultaneously in thousands of lakes around the whole world due to similar unsustainable practices, then consequences such as the loss of carbon sinks, which will affect the global climate system, and the collapse of local economies could indeed become a global concern.

The third group of boundaries consists of two human-created threats: air pollution from soot (black carbon), nitrates, sulfates and other particles; and pollution of the biosphere by chemicals such as heavy metals and persistent organic pollutants. Because of the hazards they pose to human health and to the

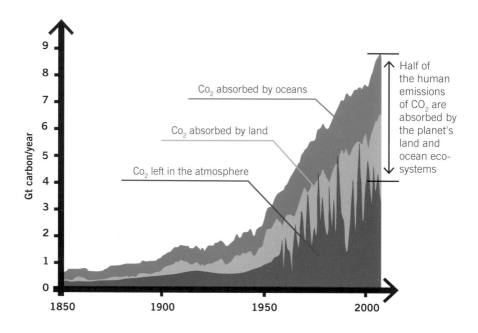

Figure 2.2 CO$_2$ Absorbed by Earth. During the past 50 years, global CO$_2$ emissions (in billion tons of carbon emissions per year) have roughly doubled. Because CO$_2$ is a long-lived greenhouse gas that stays in the atmosphere for up to 1,000 years, one would expect that the entire amount of carbon dioxide emitted has contributed to the 1°C (1.8°F) warming observed during this period. However, the reality is that the ocean and land ecosystems of the biosphere each absorb roughly 25 percent of our emissions, leaving only half of the total in the atmosphere (the clouded area in the graph). This means that in the past half century, the amount of CO$_2$ absorbed by nature has increased from 2 billion tons to 4 billion tons—proof of Earth resilience at play.

Earth system at local, regional, and global scales, our scientific team felt that both problems merited planetary boundaries. But because both are associated with a multitude of processes in a complex way, we haven't yet managed to set safe boundaries for them. More research is needed.

PLANETARY BOUNDARY UPDATES

Since our original study, we've made two key advancements in the way we frame the planetary boundaries. The first was to recognize that there is a certain hierarchy among boundary processes. It is one thing to distinguish them based on whether we have evidence of planetary-scale tipping points (The Big Three), or whether the processes operate as resilience regulators (the slow variables). But an equally important question is whether some boundaries on their own can knock the Earth system away from the Holocene, or whether they only regulate the outcome of others?

Our analysis indicates that the climate system and the richness of biodiversity on Earth have a decisive role, on their own, in determining the outcome of the planetary state. The final state of our climate system and biodiversity is determined by the aggregate effects of how water flows, land use, and nutrient flows operate. In simple terms, if we get it right on climate and biodiversity, then we're very likely to safeguard a desired state of the planet. We therefore call these two "core boundaries." Conversely, the only way to succeed in ensuring that we stay within a safe operating space for climate and biodiversity, is to get it right on all the other boundaries, since these determine the life conditions for all species in the biosphere and regulate the final state of the climate system.

The second advancement of the boundaries framework resulted from a critique of the original proposal. In that proposal, we defined only global boundaries, which, as many rightly have pointed out, provided little guidance for "managing the boundaries" at the local and regional scales. The fact is, all of the processes operate locally: Emissions of greenhouse gases have local sources, just as land use, freshwater use, and deforestation all occur locally, even though they have global consequences. It makes sense, therefore, to couple the global boundaries with regional-scale boundary definitions, since the boundaries operate across scales. Each planetary boundary, where relevant and possible, now has a twin definition—a planetary scale and a regional scale boundary level. With respect to freshwater, for example, we have defined the global boundary as the maximum amount of global runoff we can use (less than 4,000 km³/yr). To this we have added a boundary for each river basin in the world, defined as the minimum amount of water that needs to be retained in each river

in order to safeguard water-dependent ecosystem functions and resilience.

But back briefly to the core approach, and how we defined boundary levels. For each boundary process where we have enough scientific evidence, our team of scientists identified a parameter that determines or describes how well, or how poorly, each process is working. We called these measurable indicators "control variables." The choice of a control variable for each boundary was based on our assessment of which indicator would provide the most comprehensive, aggregated, and quantifiable description of how the boundary process is operating. In the next chapter, we'll dive deeper into the most urgent of these boundaries and the global risks they pose. But for now, here's a brief description of how we proposed to track them.

For climate change, as we mentioned earlier, we suggested two control variables: atmospheric concentration of CO_2 and radiative forcing. On the one hand, we proposed a boundary of no more than 350 ppm for atmospheric CO_2 concentration, based on the fact that paleo-climatic data show that concentrations higher than that can result in the melting of the large polar ice sheets, with potentially disastrous consequences for humanity, from rising sea levels to the collapse of coral reef systems and rainforests. At the same time, it was necessary to complement this "simple" carbon dioxide boundary with a boundary that incorporates the effects of all greenhouse gases, both warming (methane and nitrous oxide) and cooling aerosols (nitrate and sulfate). The best parameter here is radiative forcing, which measures the net amount of energy at Earth's surface added by human emissions of greenhouse gases. We proposed an increase of no more than 1 watt per square meter. This is generally understood to correspond to a rise in average global temperature of 1°C (1.8°F). Right now, these two variables closely follow one another, because the warming by all other gases other than CO_2 (roughly 20 percent of the forcing so far) is cancelled out by the net effect of all cooling gases. But given the complexity of the atmosphere, including the interplay between greenhouse gases and cooling pollutants, this might not always be the case. So we felt it was prudent to measure both CO_2 concentration and radiative forcing as markers for climate change.

We took a similarly pragmatic approach when choosing control variables for other planetary boundaries. For ocean acidification we proposed measuring the surface seawater saturation level of aragonite as a proxy indicator, since it's a good indicator of overall ocean health with regards to ocean acidification.

Previous pages: Small parcels of land in Rwanda's Gishwati Forest Reserve (left) were given to returning refugees. In Sarawak (right) urban expansion crowds former coastal rainforest.

Aragonite is a type of calcium carbonate that dissolves when ocean water becomes too acidic. If the aragonite level were to drop below 80 percent of pre-industrial levels, we estimated, then corals reefs could also be endangered, potentially leading to a collapse of critical marine ecosystems.

With respect to the rate of biodiversity loss, we proposed to use the current extinction rate of species as a parameter of overall ecological impact. Studies of the fossil record have shown that the average extinction rate for both marine organisms and mammals in the past was 0.1–1 extinctions per million species per year. Since the advent of the Anthropocene, that rate has skyrocketed to more than 100 extinctions per million species per year—up to a 1,000-fold increase over the geological background rate. Today about a fourth of all species, among those that are well studied, are threatened with extinction.

Scientists have warned that if species loss continues at that scale it could undermine the ability of many ecosystems to keep functioning, which would be bad news for human societies that depend on them. It could even cause these ecosystems to tip into undesired states, such as savannahs turning into deserts, or rainforests into savannahs. But defining a safe level has remained extremely difficult. For one thing, every species isn't equally important in the functioning of ecosystems. The loss of a top predator species such as sharks can be much more damaging to a reef system than the loss of other fishes. So, given that uncertainty, we proposed as an interim indicator a maximum of ten cases of extinction per million species per year—significantly higher than the background rate but one to two orders of magnitude less than the current rate. We've improved the biodiversity indicator further, by complementing the rate of species loss with an indicator that measures the capacity of ecosystems to deal with change. It does this by considering not only the number of species but also the different functions that species play, such as pollination and whether fish are predators or grazers. It also considers the number of species for each function that still exist in an ecosystem (as measured by indices such as the Means Species Abundance or the so-called Biodiversity Intactness Index, or BII).

For stratospheric ozone depletion, which poses health hazards to humans in the high latitudes from damaging ultraviolet (UV) rays, we proposed a boundary of no more than a 5 percent loss in O_3 concentration from pre-industrial levels. Greater losses than that could increase the likelihood of ozone "holes" reappearing over the polar regions each spring.

For global freshwater use, we proposed a boundary of no more than 4,000 km³ per year of consumptive use of runoff resources. That would essentially mean a leveling off in human exploitation of rivers and aquifers for irrigation and other uses to avoid thresholds that could lead to the collapse of terrestrial or aquatic ecosystems.

Conversion of rainforests and other ecosystems to cropland and cities represented another problem calling for a planetary boundary. To keep from crossing dangerous thresholds such as biodiversity loss, freshwater disruption, and reduction of carbon sinks, we proposed a boundary of no more than 15 percent of the planet's ice-free land surface under cultivation or development—compared to the current figure of about 12 percent. We chose this control variable as it is well known and monitored at the global scale. We recognized that it was a rough simplification, as our real concern for land is to make sure that we maintain the most critical natural land-use types intact, such as the remaining forests, which play a key role in regulating carbon sinks, constitute habitats for biodiversity, and regulate water flows across regional scales. In our latest update we have therefore switched the boundary definition from one side of the coin (the maximum amount of crop land added) to the other side of the same coin (the minimum amount of forest areas necessary to maintain critical biomes). Our latest proposed assessment is that we need to maintain intact at least 85 percent of rainforests (in Amazonia, the Congo Basin, and Southeast Asia), 85 percent of the world's boreal forests, and 50 percent of the world's temperate forests.

With respect to the overuse of nitrogen and phosphorus, which threaten to eutrify marine ecosystems and trigger major anoxic events or "dead zones," we proposed a boundary of no more than 35 million tons (or Mt) of N_2 per year of industrial and agricultural fixation and no more than ten times the natural background weathering of phosphorus in the annual inflow to the world's oceans. In our latest update we have improved the nitrogen boundary by including not only artificial nitrogen fixation by the Haber-Bosch process of producing fertilizers, but also the added biological fixation of nitrogen in modern agricultural systems (nitrogen uptake that would not occur if it had not been for our modern agricultural systems). This has led to an updated nitrogen boundary of no more than 44 Mt of nonreactive nitrogen gas extracted from the atmosphere and transformed to reactive nitrogen in the biosphere per year. Furthermore, for phosphorus, we were criticized, in our original analysis, for considering only catastrophic thresholds in the ocean. As Steve Carpenter and Elena Bennett clearly showed in a critical update on the phosphorus boundary, well before phosphorus causes damage downstream on marine systems, it causes tipping points in freshwater systems. Phosphorus leaching originates not only from land, particularly from fertilizer use in agriculture, but also from weathering and leakage from water-treatment plants. As phosphorus moves through landscapes, from land to the sea, it triggers thresholds in lakes and wetlands. We've therefore added a freshwater phosphorus boundary to the ocean phosphorus boundary.

By quantifying planetary boundaries in this way—as clearly marked guardrails for human impacts—our aim was to advance integrated science for global sustainability and provide policymakers, business leaders, and members of the public with practical tools to keep us from skidding off dangerous cliffs. (For more information about the nine planetary boundaries, their respective control variables, and the potential consequences of transgressing each, see tables on pages 50 and 61.) Viewed from a slightly different angle, though, the planetary boundaries also offer a hopeful, positive promise, laying out a safe path for humanity that gives us plenty of operating space for innovation, growth, learning, experimentation, and diversity. If we stay on the safe side of planetary boundaries, that is, we can continue to develop and prosper for many, many decades and centuries to come.

WHERE DO WE GO FROM HERE?

The purpose of our research, as we've said, is to map out a safe operating space for humanity in order to avoid crossing sudden catastrophic tipping points in the Earth system. Once a boundary is transgressed we enter a danger zone where crossing thresholds can no longer be excluded. As the data clearly show—see the chart on page 65—we've transgressed four of the nine planetary boundaries and are thus in a danger zone already: climate change, biodiversity loss, global land cover change, and nitrogen overload and the freshwater part of the phosphorus boundary.

The most recent observations of CO_2 levels in the atmosphere show a monthly average concentration of 399 ppm, well beyond our planetary boundary of 350 ppm. In fact, the monthly CO_2 average hasn't been as low as 350 ppm since 1986, and it has climbed steadily every year by an average of 1.4 ppm. Although greenhouse gas emissions in the USA and other nations have recently showed modest reductions, the trend is still strongly in the wrong direction. In fact, despite US and Chinese promises in November 2014 to peak CO_2 emissions by 2030, and the EU's reduction target of 40 percent by 2020 (compared to 1990), the world continues to rush toward the 450 ppm CO_2 boundary ceiling, beyond which catastrophic tipping points are, from a scientific perspective, very likely.

The same is true of the vast quantities of nitrogen being released into the biosphere, primarily through commercial fertilizers for modern agriculture. Excessive nitrogen loading in lakes, rivers, and wetlands threatens to trigger tipping points on local to regional levels, such as those we've seen in the Baltic Sea, where about a sixth of the water is now a "dead zone" with low-oxygen content. The current scientific assessment about the global nitrogen cycle is that a safe planetary boundary should be the production of no more than 44 Mt of nitrogen per year, a level we raced past in the early 1990s. To reach that level from the current one of about 150 Mt per year would require cutting back on more than two thirds of nitrogen production. Another tall order.

But the most dramatic transgression is what's happening with the biodiversity boundary. The pace of species loss today is so great we're literally in the midst of the planet's sixth mass extinction, which is certain to cause massive and permanent changes in the functioning of Earth's ecosystems. Especially troubling are the losses of top predators, species at the top of the food webs, which are rapidly changing the entire structure of natural life-support systems, including triggering major tipping points. These losses are tragic, because, unlike other planetary boundaries, they're irreversible. A lost species cannot be brought back. That makes biodiversity loss a deep global concern.

Our most recent update shows that we are in a danger zone also on land-use change. We have cut down so much of the tropical rainforests and temperate and boreal forests that only about 60 percent of Earth's original forest cover remains. Our estimate is that we need to keep at least 75 percent of Earth's forest cover to safeguard the planet's resilience.

Although we are still operating within a safe space in relation to the other four planetary boundaries (we cannot know for sure for novel entities), we see the same disturbing trend of rapid growth for all of them. This applies in particular for freshwater, where about a fourth of the world's rivers no longer reach the oceans, and there have already been several freshwater-driven collapses at the regional scale, such as Lake Chad and the Aral Sea. The trend is looking very gloomy indeed.

One additional complicating factor is the way that planetary boundaries interfere with one another. If we transgress one, we can't exclude the possibility that we'll cause others to shift, most likely in a direction that further constrains our options. Rather than being permanently fixed goals, in other words, planetary boundaries are dynamic targets, constantly moving, which means they can't be managed one at a time. That's an important message for policymakers, investors, business leaders, scientists, and concerned members of the public. Our advice therefore is to think of the planetary boundaries as a comprehensive package that operates a bit like the "three musketeers:" one-for-all and all-for-one. To avoid slipping away on any single boundary, we need to stay within the safe operating space for all of them. Conversely, if one boundary is transgressed, such as climate change, it's more likely that we'll lose on others as well, such as biodiversity.

It's becoming increasingly clear, in the end, that the "final battleground" over climate change is moving away from a focus on reducing emissions to a focus on managing the biosphere. Whether we're able to achieve a low- or even zero-carbon society within 50 years (as challenging as that may be) won't be the only factor determining whether we go over the climate cliff. We must also ensure that the biosphere, with its boundaries on land, water, biodiversity, and nutrients, continues to provide enough resilience to buffer global warming by sequestering carbon, keeping methane below ground, providing freshwater for biomass growth (which locks carbon), and maintaining resilient habitats of grasslands, savannahs, rainforests, and wetlands. If we transgress boundaries for all these ecosystem functions and services, we can't count on the planet behaving as we want it to. We can't expect it to maintain the negative feedbacks that have so far slowed global warming. If the planet goes from friend to foe, then our mitigation policies for emission reductions will make little difference. Because as impressive as they might be, human impacts on the climate system pale by comparison to the warming feedbacks the Earth system itself can trigger. If the carbon contained in just the top 50 cm (19.6 inches) of soil in the Arctic were to be released, for example, it would exceed all the carbon humans have emitted since the industrial revolution began 250 years ago.

We used to live in a small world on a big planet. Now we inhabit a big world—with big impacts—on a small planet.

That's why we need to respect the planetary boundaries.

3

BIG WHAMMIES

SOMETHING SURPRISING happened to Greenland's massive ice sheet in July 2012. One glaciologist described it as the "nightmare" they'd been dreading: For the first time in observed history, the entire ice sheet was melting.

Ice always melts along Greenland's coast during the summer. To witness the Ilulissat Icefjord breaking off massive chunks of ice into the sea at this time of year is truly overwhelming. As pieces calve off the ice sheet, they form a seemingly endless archipelago of icebergs floating down the west coast.

But over the course of four days in July 2012, a persistent high-pressure system locked warm air over this vast ice sheet mass, turning a thin layer into slush across almost the entire ice sheet (97 percent) for the first time in observed history. As a result, its surface shifted from white to a darker color, with dramatic consequences for the atmosphere.

Ordinarily during the summer, the ice sheet's bright surface reflects 85 percent of incoming heat from sunlight back to space. But during this short period of unprecedented surface melting, the ice sheet's darker surface absorbed more than 50 percent of that heat. That tipped Greenland from a net "cooler" of Earth's atmosphere to a net "heater."

In fact, Jason Box and his team at the Byrd Polar Research Center at Ohio State University, estimated that during this extraordinary two-week period Greenland injected an estimated 300 exajoules (EJ) (10 followed by 18 zeros) of energy into the atmosphere. To put this into perspective, the annual energy use of the entire world is approximately 600 exajoules. (The USA, the largest energy-using nation, consumes some 200 exajoules.) This means that, for a short time that summer, Denmark surpassed China and the USA as the country having the greatest impact on Earth's climate—not because the Danish people were dumping more greenhouse gases into the atmosphere, but because the ice sheet on Greenland stopped reflecting all that energy back into space.

When lakes and rivers are overloaded with nutrients and chemicals from urban and agricultural runoff, algae blooms can choke the ecosystem by consuming all the oxygen when they decay.

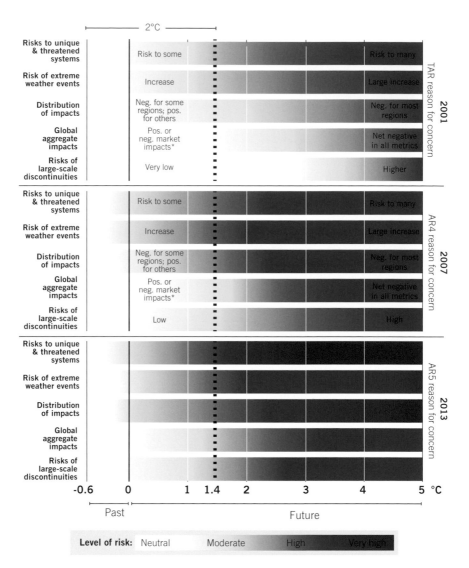

Figure 3.1 Rising Global Risks. The more we learn about how the Earth system works, the larger our reasons for concern. As climate science advances, the level of risk has risen, here shown as "red embers" of risk from the past three assessments of the Intergovernmental Panel on Climate Change (IPCC), the third assessment (TAR) in 2001 at top, the 2007 fourth assessment (AR4) in the middle, and the latest fifth assessment (AR5) from 2013 at bottom. As seen from the graph, the risk for climate-induced disasters (the last category in each assessment) was estimated at 4–5°C in 2001, but as low as 2–3°C in 2013. The 2°C (3.6°F) target for global warming is shown by the dashed line, including past (1900–2000) warming and future warming.

* Majority of people adversely affected

This was just an early warning. We don't know yet if Greenland has actually passed a tipping point, beyond which the island will enter a phase of permanent self-reinforced warming. The ice sheet hasn't shown the same dramatic melting since 2012. The tricky thing is that if Greenland does this a few times, it could induce a self-reinforced warming that will be impossible for us to stop. As Box told one reporter, "the sleeping giant is awakening."

The Greenland ice sheet contains enough water that, if all the ice were to melt, it would raise global sea levels by about 7 m (23 ft), with catastrophic impacts on coastal cities and regions. But as the 2012 "flash" melting demonstrated, the biggest danger of transgressing planetary boundaries isn't so much that such an event will cause an immediate crash. It could take hundreds or even thousands of years to melt Greenland's entire ice sheet. Rather, it's that such a disturbance could light the fuse on a planetary "time bomb" by triggering changes in feedback processes (from "negative," dampening processes to "positive," self-reinforcing processes) in which Earth takes over, transforming the initial event into a self-accelerating, irreversible engine of change so strong it pushes the planet into another state.

Should that happen, as mentioned in Chapter 1, Earth could turn from friend to foe. Instead of counteracting human pressures through negative feedbacks, Earth could launch runaway positive feedbacks with enormous consequences. And there'd be nothing we could do about it.

The fact that ice and snow, because of their white surfaces, reflect incoming heat from the sun back into space, is one of the most important and well-known of Earth's negative feedbacks. By helping to cool the planet, this process also helps keep Earth in its current stable state. But as the atmosphere warms, more ice and snow vanishes—faster than predicted. Between 2004 and 2008, the Arctic Ocean lost 42 percent of its multiyear sea ice, something experts hadn't expected until 2030 at the earliest. During the summer of 2007 alone, the Arctic Ocean lost 30 percent of its seasonal sea ice cover. While the rest of the world is still adjusting to the realities of 1°C (1.8°F) of planetary warming, the Arctic has already been experiencing a 2°C (3.6°F) world.

The question is: Has a fuse already been lit on a planetary-scale climate bomb?

ANTARCTICA: WEAK BIG BROTHER?

Earth has gone through warm periods before. About 120,000 years ago, during the so-called Eemian period, temperatures in Greenland were at least 4°C (7.2°F) higher than they are today, sometimes surging as much as 8–10°C (14.4–18°F) higher. Thanks to studies of marine fossils along coastal shore lines, we know with a high degree of certainty that global sea levels during this period were approximately 4–8 m (13–26 ft) higher than they are today. Melting ice sheets were a dominant cause, presumably with a large contribution from Greenland.

A recent analysis of Greenland ice cores, however, tells an unexpected story. When Dorthe Dahl-Jensen of the Niels Bohr Institute in Copenhagen and her colleagues on the North Greenland Eemian Ice Drilling project (NEEM) studied ice cores from 120,000 years ago, they found that the Greenland ice sheet had not melted as much as scientists would have expected. In fact, during this prolonged and great warming, the 2.5-km (1.5-mi) thick ice sheet lost only about 400 m (0.25 mi) of ice. It appears that Greenland is more resilient to a warming shock than previously suspected. At first blush this sounds like good news!

But that raised a troubling question. If Greenland had lost only 400 m of ice, an amount estimated to have contributed about 2 m (6.5 ft) to the global sea level rise, where did the rest of the water come from? Where are the 2–6 m (6.5–19.5 ft) missing from the equation? There's really only one possible answer: Antarctica.

For researchers, it was always the Arctic, not the Antarctic, which was seen as the vulnerable pole. Antarctica has always been considered the resilient big brother. The massive amounts of water in the ice sheets there—enough to raise global sea levels by some 70 m (230 ft) if they melted—always seemed safely locked away. But perhaps we are mistaken. Maybe Antarctica isn't as resilient as we have always thought.

Two recent reports by independent research teams appear to suggest just that. They conclude, based on observations, that the Thwaites Glacier, a keystone holding the massive West Antarctic ice sheet together, and several neighboring and interconnected glaciers, may have irreversibly started to melt. Since 2007 these glaciers have poured as much meltwater into the Amundsen Sea in West Antarctica as all of Greenland has in the north, some 280 billion tons a year. And that amount has risen rapidly since then.

For West Antarctica, this represents a worrying tipping point, as warming ocean water melts these massive ice sheets from below. This in turn lubricates the sloping bedrock on which they rest, setting them irreversibly in motion, slowly slipping into the sea. The only thing holding the entire Thwaites Glacier from collapsing into the sea, at this point, is a 600-m (0.37-mi) deep ridge on the sea floor. Researchers believe that the Thwaites Glacier functions as a massive ice plug, holding the other glaciers in West Antarctica in place. If this plug were to be pulled out, the other glaciers could also slide into the sea.

We won't see a gigantic domino-like tumbling of glaciers all at once, of course. It's more likely to take place over the next 200–500 years, committing the world to at least 3 m (9.8 ft) of additional sea level rise. This may seem like a slow pace and a long time period. But the fact is, the moment of truth is right now. The fate of Antarctica will be determined by whether we push the "on" or "off" button today. Once a tipping point is crossed, it will be too late. We actually write the future in the present.

Even if we take actions immediately to slow the pace of melting in Antarctica, however, we may already be committed to an additional 1 m (3.3 ft) of global sea level rise during this century. This is in addition to the 1 meter already estimated by the IPCC for this same period. The fact is, we don't know how to adapt to such a pace. As glaciologist Richard Alley of Pennsylvania State University pointed out when these studies were released, crossing this tipping point in Antarctica means that we are now committed to a global sea level rise equivalent to a permanent Hurricane Sandy-size storm surge.

MESSAGE FROM EARTH: PAYMENT DUE

Until now, the remarkable resilience of the Earth system has given human development a "free ride." Even during the current era of massive abuse of the planet—especially since the great acceleration of pressures that began in the mid-1950s—the planet has been remarkably forgiving, applying dampening (negative) feedbacks to absorb most human impacts. Despite our business-as-usual practices—emitting greenhouse gases, losing biodiversity, polluting air and water, over-extracting natural resources, degrading land and forests—Earth has pushed ahead without major repercussions.

In fact, the model of economic growth at the expense of the environment model has actually worked quite well! As Al Gore has pointed out, we've profited handsomely from using our climate as a sewer. For those nations that took part in the industrial revolution, this approach generated immense wealth extremely quickly. It worked because the growth was subsidized by Earth: More energy, more resource use, greater consumption of ecosystems, largely for free.

But the era of "massive Earth abuse" is now over. We've reached the end of the road with our current development paradigm. Not because we've run out of resources. Nor because of poor water, polluted air, or deteriorating ecosystems. But, rather, because we're approaching a point where the pressure we put on the planet may push the wrong "on" buttons, triggering Earth to kick in with self-reinforcing "positive" feedbacks—like the rapid warming of Greenland from loss of reflective snow and ice. These are the "big whammies" we need to avoid above all else.

Earth's message is loud and clear: It's time to pay up. Red warning flags are flying everywhere. Humanity has already transgressed three of the nine planetary boundaries our team of scientists identified in 2009 to keep the world in a safe operating space: climate change, rate of biodiversity loss, and the global nitrogen cycle. Other thresholds are also in danger of being crossed, such as land-use change (because of deforestation and urban expansion) and freshwater use (because of the ever-increasing demand for food, which requires huge amounts of freshwater during its production). Recent analyses by our scientific peers show that the global phosphorus cycle has also reached a global danger zone.

Consider the situation with climate change. The year 2014 will almost certainly be remembered as a milestone in global climate risk. This was the year when CO_2 concentrations in the atmosphere reached the critical level of 400 ppm—450 ppm for all greenhouse gases. Political leaders have generally accepted the idea that a greenhouse gas concentration of 450 ppm will correspond, on average, with a 2°C (3.6°F) increase in global warming. As many scientists have shown (and our planetary boundaries research strongly supports), this is a very optimistic assumption. In fact, most risk analyses indicate that we must stabilize the atmosphere at a lower GHG concentration to avoid a 2°C rise in global temperature. But be that as it may, a rise of 2°C is agreed upon as the boundary beyond which we could cross catastrophic tipping points.

Now we've reached that ceiling. We've saturated the atmosphere. Yet we still haven't changed our ways. As Corinne Le Quéré and her large international team of Future Earth scientists in the Global Carbon Project have shown, global emissions of carbon actually surged in 2014. Their estimates show that, in 2013, a staggering 36 billion tons of CO_2 were emitted—the most in human history! And projections show that we will likely break through the 40-billion ton barrier by 2015. In fact, we're following a path that, on average, takes us to 4°C of warming by the end of this century.

This is unacceptable, in our view. We believe that even 2°C of warming will lead us into very dangerous territory. The last time we had 2°C warming—120,000 years ago—the global sea level was 4–8 m (13–26 ft) higher than today. Based on everything we know about how Earth has fared in its geological past, 4°C of warming will create nothing less than a global crisis in which we have no idea how to feed humanity, much less preserve cities like New York or Sydney.

Previous pages: A farmer in India's Madhya Pradesh state clears land by burning the forest. Slash-and-burn methods cause health problems and regional-scale air pollution that affects rainfall patterns.

As if that weren't enough, we must always remember that these estimates could be overly conservative, since they're based on optimistic assumptions about Earth's continued resilience (from large carbon sinks in the biosphere) and no tipping points. In the face of potentially catastrophic outcomes, we need to consider the probabilities of even the improbable.

Is what we have to look forward to: A gradually less beautiful world? A gradually weaker planet? A more expensive place to live, where resources such as oil and metals are harder to find, and ecosystem services such as food and drinking water are more scarce? A place where healthy conditions like air quality are more difficult to deliver? Is this the price we have to pay for using up the safe operating space on Earth?

BIODIVERSITY ON THE BRINK

The risk of triggering "big whammies"—of crossing thresholds that bring abrupt planetary-scale changes—is not unique to the climate system. We know from vast empirical evidence that many ecosystems, from local lakes to forests and coral reefs, also have tipping points. After remaining in a stable state for a long time, such ecosystems can flip abruptly into another state.

Take a rainforest, for example. Under pressure from deforestation and climate change, a rainforest can abruptly shift into a savannah, and get locked in that new stable state. To maintain a stable state, an ecosystem needs feedbacks that reinforce that state. For a rainforest, the feedback that keeps it stable is the self-generation of moisture and rainfall, thanks to its vast canopy. But when a rainforest is opened up by the cutting of trees, and the atmosphere gets warmer, the system gradually dries out. Its resilience is lost. Eventually it gets to the point where the system crosses a threshold, and the feedback changes direction, from self-generating moisture to self-generating dryness. Suddenly dry air flows through the opened canopy, evaporating moisture that was previously held within the system. Less rain is generated due to less pumping of water from tree roots. The system becomes self drying, locked in a savannah state.

Two other examples of ecosystem tipping points that we've tracked at the Stockholm Resilience Centre include hard coral reef systems, which can suddenly collapse under the pressure of ocean warming, nutrient overload, and overfishing, leading to the establishment of soft coral systems or reefs dominated by seaweed; and lakes, wetlands, rivers, and groundwater systems, which can suddenly explode with algae blooms and anoxic conditions when runoff from farms and urban areas overloads them with nitrogen or phosphorus.

Of all the factors contributing to the resilience of an ecosystem, and thus helping to maintain its current state, perhaps the most important is

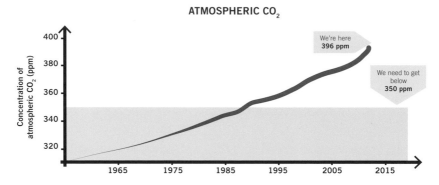

ATMOSPHERIC CO$_2$

Figure 3.2 Atmospheric CO$_2$. The planetary boundary for climate is set at 350 ppm CO$_2$, based on an analysis of the sensitivity of the climate system to increased greenhouse gas concentrations in the atmosphere, on the behavior of the large polar ice sheets under conditions warmer than those of the present geological epoch, and on the observed behavior of the climate system at a current CO$_2$ concentration very close to 400 ppm. As the chart shows, we are already beyond the climate change boundary and in the danger zone. The challenge before us, therefore, is not just to change the trajectory of our path of an ever-increasing concentration of CO$_2$, but also to remove CO$_2$ from the atmosphere if we are to move back into a safe operating space for the climate.

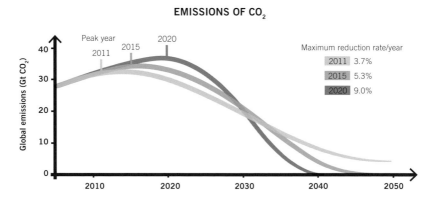

EMISSIONS OF CO$_2$

Figure 3.3 Staying with a Global Climate Boundary Means Operating Within a Finite Carbon Budget. A world transformation into the safe operating space of the climate boundary will require the decarbonization of the world economy by mid-century. The earlier we bend the global CO$_2$ emission curve, the easier it will be to return into the safe operating space. Waiting until 2020 before bending the curve will impose an almost impossibly steep 9 percent per year rate of global reductions.

biodiversity. When top predators like sharks, wolves, lions, or cod—or critical grazers like parrot fish or surgeon fish—are removed from an ecosystem, the entire food web can be thrown out of balance, triggering tipping points that can abruptly push these ecosystems into a different state. The same thing is true of Earth. Biodiversity operates from the "bottom up" as a planetary boundary that regulates local ecosystems, becoming a global concern if enough systems topple at the same time. The stability of the whole planet relies on a myriad of stable ecosystems, which in turn depend on the richness of different functional groups of species—from bacteria in the soils to pollinators and top predators.

Since we published our analyses of planetary boundaries in 2009, new evidence has come to light suggesting that biodiversity can also function as a "top down" planetary-scale tipping point. A recent analysis by an international group of ecologists led by Anthony Barnosky of the University of California, Berkeley, recently presented evidence that by mid-century we could face a planetary-scale tipping point if we continue to lose biodiversity at the current pace. The combined pressures of population growth, widespread destruction of natural ecosystems, and climate change may be driving Earth's biosphere toward an irreversible change. This could lead to a collapse of many, if not most, agricultural systems on Earth, which depend on a balanced configuration of species, including microbiota for soil productivity, and pollinators for seed and fruit development. Without adequate preparation, such a planet-wide tipping point would have massively destructive consequences—one of them being to undermine our ability to feed ourselves.

CORAL REEFS: THE "CANARIES" IN THE CLIMATE COAL MINE

If you want to know what's happening in the oceans, pay close attention to the world's major coral reefs. Often described as the "rainforests of the sea," coral reefs host a wealth of biologically rich and productive ecosystems. They could also be called the "canaries in the coal mine" of climate change, because they're so sensitive to shifting conditions. Whenever there's a change in the ocean, it often shows up first in coral reefs.

Lately, the coral reefs have been suffering badly. Even the richest coral reefs on Earth, such as those off the Raja Ampat Islands in Indonesia's West Papua province, haven't been immune to the impacts of global changes. Even here, far from pollution and overfishing, warming seas have triggered coral bleaching, which occurs when corals lose the micro-algae that live within their tissues and provide their lively colors.

It only takes a few weeks of warmer-than-usual seas—from 1–2°C (1.8–3.6°F) above normal—for widespread bleaching to occur. As we know, 95 percent of

the heat increase caused by greenhouse gas emissions is trapped in the oceans, making climate change a key trigger behind the rising severity of bleaching events. Moreover, as Daniel F. Gleason and Gerard M. Wellington of the University of Houston reported two decades ago, bleaching often occurs in combination with higher levels of solar radiation. In addition to higher water temperatures, many of the reefs in the Caribbean Sea, where bleaching was observed by Gleason and Wellington, had also been exposed to higher-than-average intensities of UV-radiation. What caused the increase in UV-radiation? It was probably a thinning of the stratosphere's protective layer of ozone, due to the presence of chlorofluorocarbons (CFCs) and other chemicals emitted by industries and communities.

Another problem has emerged in the Great Barrier Reef of Australia. During the past 20 years or so, healthy calcification by coral seems to have decreased by 14 percent. Although research suggests that increasing water temperatures provided some of the stress affecting the coral, science shows that another factor is at work, making it difficult for reef corals to build their calcium carbonate skeletons. That factor is increased ocean acidification.

Almost a third of the CO_2 pumped into the atmosphere by humanity dissolves in the oceans. There the CO_2 forms carbonic acid, which decreases the pH of seawater, making it more acidic. This in turn reduces the concentrations of carbonate ions in the water, which corals need to grow their skeletons. Left unchecked, ocean acidification can even cause coral skeletons and reefs to dissolve.

Research indicates that coral reefs weakened by overfishing and pollution are less likely to survive rising water temperatures and higher ocean acidity than pristine ones like those in the Raja Ampat Islands. Overfishing of reefs by nearby communities or international fleets removes key species of plant-eaters at the same time that runoff loaded with fertilizers and other pollutants stimulates the growth of soft coral and seaweed. The combination can be overwhelming for coral reefs, flipping them into a new and unproductive state. By contrast, reef ecosystems that maintain biodiversity have a much better chance of bouncing back from the impacts of climate change. Resilient and diverse communities, in the end, are the best defense against big whammies.

OUT OF THE BLUE?

In the new era we've created, the Anthropocene, it's not enough to acknowledge that we're putting enormous pressures on Earth through the nine planetary

Previous pages: More than 75 percent of Borneo's lowland rainforests have been cleared to make way for palm oil plantations.

boundary processes we've identified. Nor is it enough to recognize that what happens at the local level directly affects what happens at the global level, and similarly that global-scale changes impact local problems. We must also accept the fact that, when tipping points are crossed, all of these cross-scale interactions can lead to unexpected outcomes.

When the EU, for example, revised fishing policies not long ago to drive high-tech fishing fleets from their "home waters," few political leaders could have anticipated that they were launching a string of events that potentially is associated with the world's worst outbreak of the ebola virus. In response to the EU's tighter fish quotas, international fishing fleets moved their operations to the coast of West Africa, where they "vacuumed" up vast stocks of fish. This was the same area where climate change, pollution, and mismanagement of local fisheries had already degraded local mangrove forests, sea grass beds, and coral reefs. The combined effect was a rapid decline in catches for African fishermen, who, faced with a shortage of food, have increasingly turned to bushmeat as a substitute to feed their families. As a result, local trading patterns shifted, with hunters killing more forest animals such as chimpanzees that are key sources of zoonotic diseases such as ebola. It's possible the current outbreak in Liberia, Sierra Leone, Senegal, Guinea, and Nigeria began in the forest when a child came into contact with the butchered meat of a wild animal infected with the virus. He may have spread the disease to others as impacts ricocheted around our interconnected world, where it is no longer possible to separate the legislative halls of the EU from the forests of West Africa.

A similarly unexpected train of events preceded the Arab Spring in 2010 2011. It began with a blistering heat wave in Russia, where massive wildfires and a prolonged drought prompted Prime Minister Vladimir Putin to restrict exports of wheat and other staple cereals. Similar actions by Prime Minister Kevin Rudd of Australia, which had suffered a dozen years' drought, combined with major speculation on the world's markets, helped to trigger a dramatic rise in world food prices. It didn't help that the global price of phosphorus, the key fertilizer in agriculture, had also risen three-fold, or that oil prices had also shot up 100 percent, boosting energy costs for farmers. As a result, food riots erupted in most capital cities in North Africa and the Horn of Africa. The region was bubbling with unrest. When Tunisian street vendor Mohamed Bouazizi set fire to himself to protest mistreatment by police, he provoked an international revolution. After decades of repressive dictatorships, a generation of frustrated young activists stood up against aging tyrants in nation after nation and the regimes began to fall like dominoes. Social unrest among common urban citizens triggered by the abrupt rise in food prices, in turn caused by Earth's

FREQUENCY OF FUTURE CORAL REEF BLEACHING EVENTS

2030

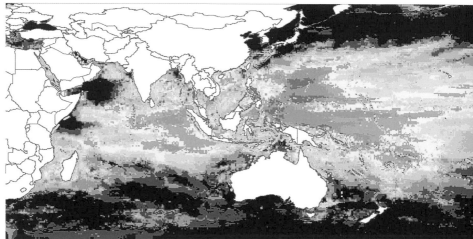

2050

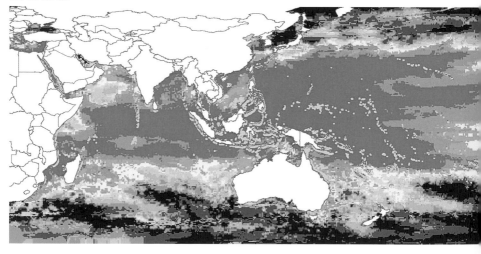

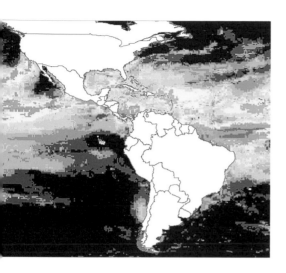

Figure 3.4 Projected Coral Reef Bleaching. This map depicts the estimated frequency of coral reef bleaching events in the 2030s and 2050s. Corals become "bleached" when water temperatures rise too high and are sustained for too long. The colors represent the percentage of years in each decade in which a National Oceanic and Atmospheric Administration (NOAA) Bleaching Alert Level 2 (severe thermal stress) is predicted to occur.

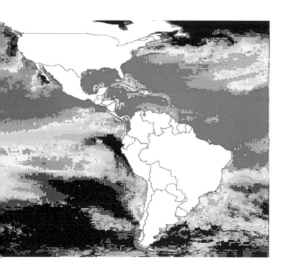

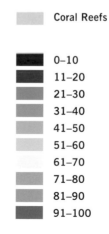

Coral Reefs

	0–10
	11–20
	21–30
	31–40
	41–50
	51–60
	61–70
	71–80
	81–90
	91–100

invoices to the world food markets, apparently interacted with the social uprising of a young generation fed up with dictatorial suppression. What had begun as a heat wave in Russia became a perfect storm of social–ecological disruption in Africa.

These kinds of long-distance interactions are a new phenomenon of the Anthropocene. Human activities in one region (like coal-fired plants in the Ohio Valley or factories in Novosibirsk) cause global environmental change (like higher temperatures from greenhouse gases), which generates a surprisingly rapid downgrading of a large Earth-regulating system in another region (like the melting of sea ice in the Arctic Ocean). How a European, American, or Chinese worker chooses to commute to the office or factory can now affect the likelihood of rainfall that will benefit farmers in the Sahel. How Southeast Asian nations manage their rainforests can now impact the frequency of heat waves in Europe or the ability of Arctic Inuit peoples to hunt on a frozen sea. These actions, in turn, cause feedbacks that further amplify global changes, creating impacts that boomerang back on the first region, or, more likely, affect some other region that may not have contributed to the original problem at all.

Because of this global web of interconnections, humanity in the Anthropocene must now consider all biomes in the biosphere—every landscape and seascape—when considering the best strategies to secure social and economic prosperity in our local communities. In a world facing big whammies, it's every nation's concern, indeed every citizen's concern, how we, as a world community, manage the entire biosphere.

This orangutan at the Nyaru Menteng Reintroduction Center in Borneo was rescued after its mother was killed by hunters.

4

PEAK EVERYTHING

THE GLOBAL CRUNCH on raw materials is getting worse every day. Within the next 50 years the world could run short of many important metals, including silver, gold, lead, zinc, tin, copper, and nickel. We may also be hitting the ceiling for economical sources of other critical resources, such as crude oil, natural gas, phosphorus, and rare earth metals. As raw materials like these get harder to find and extract, industries will be forced to devote more energy to exploiting lower-grade resources that are environmentally dirtier, putting additional stresses on modern economies and ecosystems. Even worse, if we continue to produce and consume these materials in the same unsustainable linear way that we have been—extracting them from the ground at one end of the economic process and depositing them as waste at the other end—we could drive dangerous tipping points in both the climate system and the living biosphere.

That makes raw materials a big problem.

To get an idea of what's behind the global squeeze on metals, consider the intense demand for indium, a soft, malleable, and toxic metal with a brilliant luster that is currently being consumed in unprecedented quantities to make liquid-crystal displays for flat-screen TV sets, laptop computers, and tablets. Recent estimates suggest that the worldwide consumption rate of indium has surpassed the production rate, resulting in a more than ten-fold price increase between 2006 and 2009. Even though a few optimists argue that there's still plenty of indium to be extracted from Earth's crust, the concentration is so low, in most cases, that extraction wouldn't be cost-effective and could have severe ecological implications. Others hope that a substitute might be found, but, the truth is, development and implementation could take a long time.

As the demand for indium demonstrates, our dependence on metals increases with every advance in technology. It takes about 50 different metals, for example, to produce computer chips, flat-screen TVs, cell phone screens, and other

Even in the world's most densely populated cities nature is still the source of everything.
The rising affluence of the world's population has put new pressures on natural resources.

modern conveniences. The interconnected world we now live in is an amazing place, of course, but too few of us see the connection between our digital gadgets and the raw materials necessary for their fabrication. And there are so incredibly many more consumers than there used to be. As large segments of Asia's and Latin America's populations rise into the middle classes—by one estimate as many as three billion people by 2030—their patterns of consuming metals have started to resemble those of their counterparts in the USA and Europe, moving deeper into "metal-intensive" digital and electronic lifestyles. Today there are almost as many cell phone subscriptions (6.8 billion) as there are people, and every second more than ten people buy a computer and 20 buy a cell phone. The result: a massive and rapidly growing hunger for metals. And even though these devices rapidly become more efficient, often portrayed as great sustainability successes, the sheer growth in their numbers vastly outweighs their gains in efficiency, causing a rebound effect of ever-growing planetary-scale unsustainability.

How long will economical sources of key minerals last if every human on the planet consumes even half as much as the average rich-world citizen does today? The answer is alarming. According to estimates published a few years ago, the world supply of antimony, to name one metal, could run out in only ten years. Silver could be gone in less than five. And indium could be used up in only a few years, if we don't find substitutes or start recycling more.

Tantalum, previously known as tantalium, is a hard, blue-gray, lustrous transition metal widely used as a component in alloys. As such, it is one of the metals that go into pinhead capacitors in cell phones and other electronic equipment. Sadly, mining of this rare metal has been associated with millions of deaths in the Democratic Republic of the Congo during warfare there between 1998 and 2002. The fighting coincided with a rise in the price of tantalum (caused by the growing market of cell phones), and it is widely acknowledged that one of the key motives for the civil war in the Congo was control of the country's mineral resources, including the biggest of Africa's tantalum mines.

Another source of potential conflict involves China's role in the market. Besides supplying 93 percent of the world's rare earth metals, China has also been investing in African mines and buying up high-tech waste to recover metals. What would happen if China decided to cut off the supply? In September 2010, customs officials in China refused to allow shipments of rare earth metals to be loaded aboard ships bound for Japan during a diplomatic dispute with that nation. The squabble was quickly resolved. But as consumption of rare earth metals continues to grow—the world's industries are expected to use 50 percent more rare earth metals in 2015 than they did in 2010—these kind of

HOW LONG WILL IT LAST?

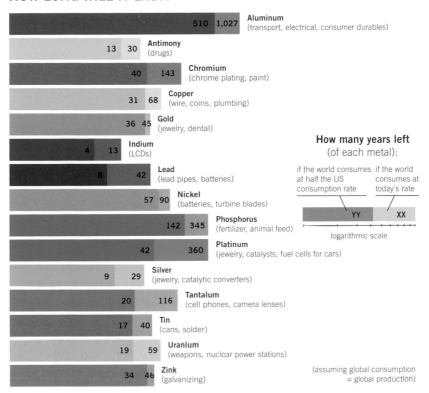

Aluminum 510 | 1,027
(transport, electrical, consumer durables)

Antimony 13 | 30
(drugs)

Chromium 40 | 143
(chrome plating, paint)

Copper 31 | 68
(wire, coins, plumbing)

Gold 36 | 45
(jewelry, dental)

Indium 4 | 13
(LCDs)

Lead 8 | 42
(lead pipes, batteries)

Nickel 57 | 90
(batteries, turbine blades)

Phosphorus 142 | 345
(fertilizer, animal feed)

Platinum 42 | 360
(jewelry, catalysts, fuel cells for cars)

Silver 9 | 29
(jewelry, catalytic converters)

Tantalum 20 | 116
(cell phones, camera lenses)

Tin 17 | 40
(cans, solder)

Uranium 19 | 59
(weapons, nuclear power stations)

Zink 34 | 46
(galvanizing)

How many years left
(of each metal):

if the world consumes | if the world
at half the US | consumes at
consumption rate | today's rate

YY | XX

logarithmic scale

(assuming global consumption
= global production)

Figure 4.1 The Global Crunch on Metals. A growing hunger for metals with which to satisfy the modern world's appetite for machines, and digital and electronic products, is exhausting accessible resources. Demand influences the velocity at which the world rushes toward peak point for many key metals. If every person in the world consumed half as much metal as an average US citizen, we would run out of easily accessible metals within the next two to three decades. This signals not only a shift toward higher and more volatile prices for metals but also the likelihood that valuable natural resources will become a significant factor in international geopolitics. The solution? We need to drastically improve efficiencies and transition into circular models for both businesses and societies.

geopolitical risks are forcing more and more governments to take the issue of peak metals seriously.

These are just a few reasons why concerns have been growing that virgin stocks of key metals may be inadequate to provide everyone on Earth with the same quality of life enjoyed by citizens in the developed world—at least given contemporary technology and business models. Manufacturers may soon need to limit development of new technologies to only those built from components that are still relatively abundant, rather than considering such factors after the fact. New business models will also be needed, such as applying the principles of a circular economy and "cradle-to-cradle" strategies, those in which the reuse of materials reclaimed from no-longer-wanted products are the engine of economic growth, rather than the extraction of ever more raw materials.

One positive development is that metals—unlike oil, coal, and phosphate—are recyclable and can be recovered from end uses, since the metals are often used in a pure form and not dissipated. The recycling of metals is likely to become increasingly lucrative as ores become more difficult to obtain, and the energy, ecological, and social costs of extracting them continue to rise. To reach sustainable levels of production, however, analysts have calculated that the percentage of metals being recycled would have to be greatly increased—from about 16 percent today for silver, 26 percent for tin, 31 percent for copper, 35 percent for nickel, 43 percent for gold, and 49 percent for aluminum—to 90 percent or greater for most metals. If industries were more efficient in their use of copper, for example, and boosted recycling rates to 95 percent, the world's economical supply of copper might last for another 600 years, instead of the 31 years currently predicted.

THE PROBLEM WITH FRACKING

Then there's the question of "peak oil." As with other raw materials, peak oil is the point in time when maximum global extraction is reached, after which the rate of global oil production enters a terminal decline. For a long time peak oil was a hugely contentious issue, much like the early controversy over climate change in the late 1980s. The debate was dominated by strong voices, generally with vested interests in maintaining a status quo, who questioned the evidence and deliberately injected manipulated "facts," disguised as science, to sow a state of uncertainty and confusion among the public.

Today, there are no longer any credible voices questioning peak oil. What

Previous pages: Vietnam. Will our urban areas become sustainable and thriving environments or implode under the pressure of rising waste, water scarcity, and pollution?

started as an assessment among scattered scientists is today a mainstream position, acknowledged both by the large oil companies and by the global institutions providing intelligence to the world from the energy sector itself. The International Energy Agency (IEA), for example, which persistently has downplayed the risk of hitting a global oil peak, acknowledged in its 2010 Global Energy Outlook that the world most likely already hit peak oil in 2006. More recently, it has argued that the exact timing of a global peak in production depends on fluctuations in demand and supply. Based on that, the IEA predicted that a peak in crude oil would prompt oil companies to turn to natural gas liquids and "unconventional oil production"— such as tar sands, shale oil, and coal-to-liquids—which still exist in abundance, though they are much dirtier and more dangerous for the climate than crude oil.

Unfortunately, that prediction has come true, which is a perfect proof why we need to define planetary boundaries within which we can have economic development and technological progress. Without boundaries, our tendency is that, with the introduction of new technologies and practices, we simply speed up our journey toward the escarpment—in this case by responding to oil scarcity by embracing alternatives that are less energy-efficient and more damaging to the climate.

Following the recent shale oil boom in North Dakota, Montana, Texas, and other states, the USA now contributes more than 10 percent of the global crude oil supply from 150,000 or so new wells. Using a technology called hydraulic fracturing, more widely known as "fracking," oil companies have spent at least 1 trillion USD during the past decade to squeeze hard to get oil out of "tight" formations as deep as 3,048 m (10,000 ft) below the surface by injecting a mixture of about a million gallons of water, sand, and chemicals into each well. The success of this effort in expanding domestic oil supplies has led some observers to claim that fracking could help revive the economy, make the USA energy-independent, and reduce energy costs for manufacturers.

But the reality is more sobering. Deposits tapped by hydraulic fracturing tend to play out much more quickly than conventional ones. Production from the average fracked well peaks after only three years, forcing oil companies to keep drilling more and more new wells just to keep up production. In addition, the profitability of fracked wells is much lower than that of conventional wells, and the environmental hazards of fracking are troubling, from the potential pollution of groundwater with chemicals to the release of methane, a potent greenhouse gas, into the atmosphere. Because of all these factors, independent analysts have predicted that the shale oil boom in the USA will fizzle out before the end of the decade.

Obviously, peak oil doesn't mean that the oil is all gone. Rather, it implies that all efforts to increase the oil production rate fail. Hence, one can no longer argue today whether or not we are running out of cheap oil. It is instead a question of how many decades of the "cheap fossil era" the world economy still has to go. The fact is, very little new oil has been found during the past three decades and the prospect of finding much more is slim. Some oil-producing regions have already experienced steep declines. From a longer perspective, it is clear that the world now needs to start a transition to renewable energy systems, not only because of resource constraints but also to preserve the resilience of our planet.

The most pressing reason to abandon the use of oil, coal, and natural gas, after all, is to reduce global emissions of greenhouse gases. If the world is to avoid dangerous impacts from transgression of the climate change boundary, we need to reduce the concentration of CO_2 in the atmosphere to no more than 350 ppm, which all credible analyses show will require nations to phase out carbon-based economies by the year 2050. In the annals of human history, in other words, the oil era itself may be remembered as a short spike in the course of the evolution of civilizations.

PEAK PHOSPHORUS

Peak oil and peak minerals aren't the only problems facing us in the 21st century. Another problem that could strike us at least as hard is a shortage of phosphorus, which could threaten agricultural production and food security for billions of people.

Modern agriculture depends on a steady supply of phosphate fertilizers produced by mining of phosphate rocks. According to some analysts, humanity is approaching or might already have passed "peak phosphorus," when global reserves of mineable phosphorus begin to decline. These studies suggest that current global reserves of phosphorus derived from phosphate rock may be depleted in 50 to 100 years. While such projections are debatable, no one doubts that phosphorus is an exhaustible resource without substitutes, which means that shortages will occur sooner or later. This is true despite the fact that phosphorus can be recycled, which tends to be difficult and costly; most ends up accumulating in the sediment beds at the bottom of our freshwater and coastal marine systems. In many parts of the world, there is already a lack of phosphorus for fertilizing fields. Of all the continents, Africa has the largest food shortage, even though it is also the world's largest exporter of phosphate rock.

At the same time that some areas are experiencing a shortage of phosphorus, overuse in other areas is the primary driver of algae blooms that cause toxicity, oxygen loss, fish kills, and other problems in lakes, reservoirs, rivers, and coastal estuaries. Most of this phosphorus comes from erosion and agricultural runoff, but some comes from human waste that is insufficiently treated. This is why, in the planetary boundaries work described in Chapter 2, we proposed a maximum amount of mined phosphorus that can be applied to Earth's soils without triggering toxic algae blooms and massive dead zones in lakes and oceans.

Phosphorus, in other words, is both a dwindling resource, a fundamental nutrient in our food, and a major pollutant, as witnessed by the fact that 80 percent of mined phosphorus is never consumed as food by humans. Instead it is wasted, causing massive damage to aquatic ecosystems, freshwater resources, and coastal areas. Today, only five countries—China, Morocco, South Africa, Jordan, and the USA—produce about 90 percent of the world's mined phosphorus. The day the world faces a peak in phosphorus supply, it will most certainly have geopolitical implications. US companies, for example, import a lot of phosphate rock from Morocco, even as Morocco is occupying Western Sahara and controls the region's phosphate rock reserves. This trading in phosphates from Western Sahara has been condemned by the UN.

To solve problems posed by imminent peak phosphorus and to avoid transgressing the planetary boundary for phosphorus, many experts believe the world should adjust its use on croplands, in food distribution and consumption, and in the treatment of human waste. On the farm, phosphorus applications should match crop needs, and erosion of phosphorus-rich soil should be minimized. Decreasing consumption of meat would be another way to decrease the waste of phosphorus. In addition, human waste contains a lot of phosphorus that could be recycled as fertilizers. The phosphorus contained in a single individual's urine, as it turns out, is roughly the same amount needed to fertilize food supply for one individual.

According to one estimate, sub-Saharan Africa could become self-sufficient in fertilizer supply if it were to adopt productive or ecological sanitation practices. This would provide the necessary supply of nutrients to smallholder farmers and provide food security and new opportunities for income. In essence, by realizing that phosphorus is a dwindling resource that must be carefully used and recycled, Africa south of the Sahara could create a win–win situation, assuring both an adequate supply of a crucial fertilizer and cleaner water. The same would also apply to richer regions like Europe and the USA.

IMAGINING A WORLD WITHOUT PEAK ANYTHING

At a time when "peak everything" is an imminent risk, resulting in a scramble for resources and price volatilities, tomorrow's winners may be those nations and businesses that protect themselves from sudden shocks by moving to closed-loop production systems, renewable energy sources, and service-oriented life cycle relations to its customers and citizens. Enterprises that remain stubbornly stuck in the current dirty, unhealthy, inefficient, and increasingly unattractive growth model, by contrast, could get left behind.

As we'll explore in greater depth in Chapter 7, we're advocating a change of perspective in this book aimed at igniting a new generation of leadership, renewal, and innovation. The world we imagine will be very different from the unsustainable and highly resource-inefficient one we inhabit today. It will have the ecological capacity and resilience to meet the aspirations of nine billion people by mid-century and eleven billion by 2100, by closing the loops of resource use, avoiding overloads of nutrients, and putting the brakes on degradation of all ecosystems as well as overuse of raw materials like rare earth metals.

The world we imagine will be efficient and modern, advancing new technologies and zero-impact systems for energy and material use. It will run a circular economy in which businesses possess a social and ecological moral compass as they mainstream "cradle-to-cradle" systems to recycle natural resources. This change of perspective means a fundamental break with the "environmentally arrogant" attitude of current linear processes in which resources are extracted and disposed of as polluting and contaminating waste. This attitude has gradually been shifting, but much more is needed, and it can only be achieved by integrating closed-loop thinking into the DNA of our societies and businesses. The "throw away" concepts must give way to systematic efforts to extend wealth and the development of circular material flows.

In such a world, there would be no need to dig, drill, or extract the world's vanishing supplies of natural resources. The very concept of a global crunch on raw materials would be obsolete.

Following pages: Coral reef fish, the livelihood for millions of small-scale coastal fishing families, dry on the beach in a village on Vamizi Island, Mozambique.

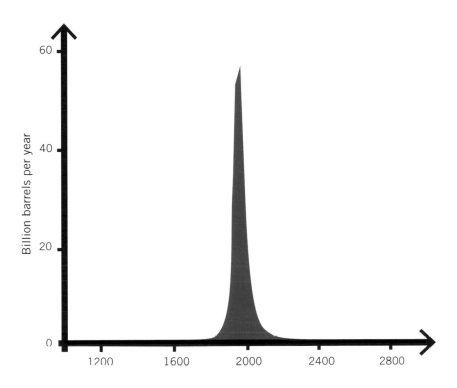

Figure 4.2 The Oil Era. Over a period of only 250 years or so, humanity has emptied deposits of oil in the Earth's crust that had taken 100 million years to accumulate. This era—during which humanity has built astonishing and unprecedented prosperity based upon cheap, easily accessible, and hugely energy-rich crude oil—appears as a historic spike.

SECTION 2

THE GREAT MIND-SHIFT

I HEAR THEM before I open my eyes. Elephant cicadas. Frogs. Owls. The birds called nightjars are still performing their nocturnal hymns. A myriad of voices take turns in the choir of the dark Borneo rainforest. Many are by now familiar, but others truly challenge the imagination. Some are melodic and beautiful, others funky and industrial, as if from a weird science fiction movie.

It's still really dark when I pull aside the tent flap and walk down the path to one of my tree blinds. I'm careful not to scare the orangutan mother and baby sleeping in their night nest up in a tree a thousand feet away. When I arrive at the giant dipterocarp tree, I put on my harness, lock into my climbing gear, and start making my way up the huge trunk. I still can't see anything except fireflies flicking through the humid air. I share the climbing rope this morning with a whole family of ants using this perfectly straight route instead of vines and lianas.

At the top, where the rope is attached to a large branch, I gently climb onto the wooden platform perched 56 m (185 ft) off the ground. My blind, covered with camouflage net and green tarp, is located opposite a fruiting giant fig tree. My hope is to get more photographs of orangutans, gibbons, hornbills, and giant squirrels. As I look through the net, the sky appears slightly lighter but I can still barely make out the leaden gray clouds flitting by.

In a few moments, the gibbons should start their bubbling calls to one another from the opposite ridge, and the argus pheasants should add their loud voices to the choir, as they've done since time immemorial. Their hooting calls, in fact, signal a healthy ecosystem. The pheasants are an important indicator species. Their presence means the forest is in good shape. When a forest is fragmented by logging, the trees become stressed and stop producing fruit or seeds. This destabilizes a system that many species depend on, and the birds disappear. A forest without the argus is a system destabilized.

I was 20 years old the first time I came to Borneo. I remember walking among the giant trees feeling as if I'd entered a cathedral. That was back in 1988. Since then I've returned 30 or 40 times, working on magazine articles, books, and documentary films. To me, this is a place of small miracles, brimming with life. No matter where you look, you see something amazing. The whip snake is one of my favorites. When it stalks a lizard in a tree, it sways gently back and forth, imitating the movement of a wind-struck liana. So beautiful.

But Borneo has changed. A lot of the island doesn't look like it used to. Much of the rainforest is gone—perhaps 75 percent—leveled to make way for plantations of palm oil trees or other crops. Even places that were supposed to be off limits have been destroyed. In some national forests, as many as 30 percent of the trees have been illegally logged—as a result of corruption.

Why should we care if we lose another rainforest? Why should we care about the orangutans or gibbons? Aren't there more important things to worry about? In a world with more than seven billion people, something has to be sacrificed, doesn't it? Why not Borneo?

This is something to ponder. After all the years I've spent working in remote and relatively pristine areas, one thing has become clear: Wild places like the Borneo rainforest are as essential to the health of faraway places like Tokyo or Chicago as the lungs in a body are to its muscles. Cities, societies, and nations depend on healthy natural ecosystems to survive and prosper. They're interconnected and interdependent.

In the chapters in this section, we'll explore this idea further, showing how sustainable use of resources, ecosystems, and the climate can help build economies. Business-as-usual isn't enough any more to deliver growth, wellbeing, or happiness. We're going to need a burst of innovative technologies, improvements in efficiency, and sustainable "circular" economies to reach our goals. We're going to have to become stewards of the remaining beauty on our planet.

Sitting up here in my tree platform, surrounded by this ancient forest, I feel small. But I also take strength from it. Borneo is one of the places that have charged my heart with passion for our natural realms. The natural world, as we know it, is not something we can choose to have or to lose. It's something we need for our survival. — *Mattias Klum*

5

NO BUSINESS ON A DEAD PLANET

THOUSANDS OF ATHLETES from around the world recently gathered in Beijing for the city's annual marathon. But on the morning of the event the air was so polluted that many runners decided not to compete. Others took the extraordinary precaution of wearing a mask to keep from getting sick. Some criticized the organizers for not canceling the event after health officials announced that the air quality index that day was "hazardous," well above the threshold at which citizens are urged not to engage in strenuous activities outdoors. Others noted that they could barely see the stadium through the smog. As one resident said, "It's really hard to breathe when it's like this."

This occurred in October 2014, but the dilemma for Beijing is that these days it's not an unusual occurrence at all. If you live in the Chinese capital, you start every morning by checking your smartphone app for air quality alerts. Regularly it's on red alert, signifying hazardous levels, which is easily confirmed by looking out the window, where thick smog makes you wonder if you're looking at a wall instead of glass.

Chinese authorities are well aware that deteriorating air quality, water quality, and food safety pose a significant threat to the nation's economic development. Bright young Chinese academics aren't attracted to live in such environments, despite the incentives of an economy with a booming GDP. The most appealing cities in the world—the ones attracting all the talent and innovation—are those offering a healthy, safe environment. It's as simple as that.

Moreover, every business leader knows that no enterprise can operate successfully in a dysfunctional society, and that the environment is a critical determinant of whether the market, the community, or the nation is functional or not. Municipal leaders in São Paulo, Brazil, for example, are struggling with water-scarcity problems caused by rising water demand and falling water supply, very likely seriously affected by both climate change and deforestation of the

As huge fleets vacuum up fish along many coastlines, fishermen like this one in Sarawak, Malaysia, have a tough time catching enough fish.

Amazon rainforest. These problems are threatening the business environment and living conditions for the city's eleven million people. Many other major cities, from Mumbai (Bombay) to Nairobi, face similar challenges with sewage, air, and chemical pollution.

It may seem obvious to point out that as we degrade our forests, waterways, soils, and air, and lose wild animals and plants, we should expect an inevitable decline in human wellbeing. We draw attention to this problem here because the scale of this decline is much greater than people imagine, with potentially serious economic consequences. As nature's resilience is lost, thresholds in Earth's physical processes are crossed that trigger runaway feedbacks, bringing abrupt changes, shocks, and stresses to our lives. As this happens, it no longer becomes possible to pursue economic development as we struggle with crisis after crisis. The future of our societies hinges on the resilience and sustainability of a stable climate and ecosystem. Or, as Petter Stordalen, CEO of Nordic Choice Hotels, says on the back of his business cards, "there is no business on a dead planet."

The overall logic is simple. A stable planet provides us with the ecosystems we've learned to love and exploit. It provides us with the functions and services we need, from clean air to healthy food. These ecosystem functions and services form the basis not only of direct wealth but also of resilience—so we're not only rich but also safe. Based on the ecological richness and the resilience that stability builds, we can fulfill our development needs and aspirations—from eradicating hunger to ensuring economic growth. But without this chain of factors, from stable ecosystems to resilience, we can't expect economic development.

You might say, well, we've had great economic growth in the past, despite environmental degradation. But until around 1990 Earth's resilience was high. Instead of incurring costs to the economy, ecosystems provided massive subsidies to the world economy. In fact, in a recent study by Robert Costanza of Australia National University and several co-authors, these subsidies were valued at about 125 trillion USD annually, or 1.5 times world GDP. The era of a free planetary ride is over.

SIGNS OF TROUBLE

One of the clearest warnings in recent times that human activities have been affecting the resilience of Earth was the collapse of the North Atlantic cod fisheries in the early 1990s off the coast of Newfoundland. After thriving for centuries under local fishing practices, the Canadian cod fishery in the late 1950s was assaulted by factory trawlers from around the world. By the end of the 1960s, the number of cod being taken increased almost four-fold to

Economic Benefits from Wise Management of Ecosystems

Biomes or ecosystems	Typical cost of restoration in USD (high scenario)	Estimated annual benefits from restoration (average scenario)	Net present value of benefit over 40 years (USD/ha)	Internal rate of return (%)	Benefit/ cost ratio
Coral reefs	542,500	129,200	1,166,000	7	2.8
Coastal ecosystems	232,700	73,900	935,400	11	4.4
Mangroves	2,880	4,290	86,900	40	26.4
Inland wetlands	33,000	14,200	171,300	12	5.4
Lakes and rivers	4,000	3,800	69,700	27	15.5
Tropical forests	3,450	7,000	148,700	50	37.3
Other forests	2,390	1,620	26,300	20	10.3
Woodland and shrubland	990	1,571	32,180	42	28.4
Grassland	260	1,010	22,600	79	75.1

Wise management of ecosystems gives economic benefits. There are ample examples today showing that when we value, even at a conservative level, the role played by ecosystems in our economic activities, then sustainable practices and businesses are more profitable than unsustainable ones. Furthermore, sustainable management of ecosystems helps to build resilience, which forms an insurance capital in the face of various shocks. This is a key dimension of economic value.

800,000 tons a year. So many cod were being hauled in, the population wasn't able to replenish itself, and by the end of the 1970s the annual catch fell to 139,000 tons. In reaction, Canada and the USA both extended their marine jurisdictions to 200 nautical miles (370 km), which pushed foreign fleets away. But foreign factory ships were soon replaced by Canadian ones, and the cod were fished to near-extinction until 1992, when Canadian federal officials finally declared a moratorium. By then it was too late. The cod fishery was destroyed.

This abrupt and irreversible collapse was a breakdown of an economic "fish-basket" worth an estimated 120 million USD in lost fish value (1989–1995) and several times more in lost incomes, ending a traditional way of life in many Newfoundland villages. More than 40,000 people lost their jobs. Similar collapses have followed in other marine systems around the world, including the Baltic Sea, where overfishing of cod and nutrient overload from cities and agriculture have turned large areas into dead zones.

Another example of human impact on the planet has come from the increased frequency of extreme weather events. Ever since we transgressed the planetary boundary for climate change—an atmospheric CO_2 concentration level of 350 ppm— in the late 1980s, we've begun to see weather disasters such as heat waves, droughts, extreme downpours, and floods more often than we used to. In 2003, almost 40,000 Europeans died as a result of the worst heat wave ever recorded in the region, the largest environmental disaster to hit Europe in modern times. In Australia, a 12-year drought from 2000 to 2012 cost the nation an estimated 4 billion USD. In Pakistan, Afghanistan, Germany, and Thailand, massive floods displaced millions of people. When Hurricane Sandy in 2012 suddenly veered inland from its expected route in the Atlantic Ocean to strike New Jersey and New York, the storm surge put Wall Street 3 m (9.8 ft) under water and cost New York City a staggering 19 billion USD. In fact, the cost of natural disasters since the 1980s has been steadily rising, and true to the Anthropocene, it hasn't been geological disasters such as earthquakes, tsunamis, or volcanic activity that have been happening more frequently, but rather climate events like heat waves, droughts, wildfires, floods, landslides, and big storms. Today, environmental disasters are estimated to cost the global economy more than 150 billion USD a year.

EARTH SUBSIDIZES WORLD ECONOMY
Another way to measure the economic impact of degrading our natural surroundings is to estimate what it would cost to replace the services we receive from them. Although we rarely consider this fact, Earth subsidizes economic

growth to an astonishing degree. In fact, if you were to take into account the deterioration of social and natural capital as a cost of economic activities, many studies indicate that GDP-based growth would be significantly lowered, if not erased. Add up the cost of unsustainable use of ecosystems, natural resources, and the climate, and economic growth largely disappears in the USA, Germany, and China.

In the 2014 study by Robert Costanza and his colleagues, humanity lost ecosystem services worth roughly 20 trillion USD a year between 2007 and 2011 as a result of degradation of the environment. To put the 20 trillion USD loss into perspective, the GDP of the entire world adds up to about 75 trillion USD a year. Costanza and his colleagues considered only "direct" services to the economy from nature, such as freshwater for food production, soil quality, timber value, and so on. They didn't include "indirect" ecological functions, such as maintaining enough top predators to ensure that landscapes remain productive and resilient or enough pollinators to keep agriculture going even in the event of a disease outbreak. For that reason, the study's estimate is conservative. Even so, it demonstrates the staggering level to which the world economy is subsidized by nature. If businesses had to pay for these ecosystem services, the result would be a 27 percent reduction in net economic output for the world economy.

There's no shortage of evidence these days that sustainable management of ecosystems is good for businesses as well as for communities, nations, and regions. In the UN's first Millennium Ecosystem Assessment (MEA) in 2005, 1,360 scientists from 95 nations estimated that almost two thirds of the critical environmental services that humans rely on today are being diminished by our activities. Similar assessments have come from the Natural Capital Project, and The Economics of Ecosystems and Biodiversity (TEEB) initiative. The Intergovernmental Platform on Biodiversity and Ecosystem Services (IPBES) has also been launched as a "sister" to the IPCC to map out the evidence of how biodiversity and ecosystem services contribute to human development. We're sitting on a mountain of knowledge showing that everything we value on Earth, from the morning dew on swaying grass to the proud herd of wildebeests trekking across the savannah, determines our ability to prosper.

Every part of the planet, living and non-living, from phytoplankton in the oceans to mineral deposits in the bedrock, interacts to establish the biophysical configuration of Earth. It's no coincidence that we have big stable biomes such as the taiga or the endlessly blue-white ice sheets, which, together with the global cycles of carbon, nitrogen, phosphorus, and water, sustain the planet as we know it. Living organisms in the biosphere, from soil bacteria to dolphins, interact to create food webs in complex predator–prey relationships. They also serve as

biological "insurance companies" by providing redundancy in the face of "low-probability, high-impact" events. In many landscapes or marine systems, different species may have the same or similar functions, such as decomposing soil or pollinating plants. But they're not equally sensitive to a virus disease or drought or shift in air temperature, which gives ecosystems what we call "response diversity"—a critical element of resilience. As nature has learned the hard way over billions of years, you don't want to put all your money on one horse.

Nature's so resilient, in fact, we've been lured into a false sense of comfort. We've convinced ourselves that even unsustainable development—losing "a few species here and there"—can't be such a big deal after all. But it is a big deal. We're in the midst of the sixth mass extinction of species on Earth. Ecosystems can cope with this, to a point. If the species we lose are lower down in the food web, they can be replaced by others that play the same functional role. But that can't go on forever. When the last critical functional group is lost, the system collapses. And as we increasingly realize, large ecosystems are particularly sensitive to losing top predators—keystone species such as cod, sharks, lions, and eagles. When these species are gone, the cascading effects can be devastating, causing entire ecosystems to crash.

WOLVES, TREES, AND BUMBLEBEES

A classic example of how change can cascade through an ecosystem took place in Yellowstone National Park. During the 1920s, wolves, one of the top predators in the park, were killed off by government hunters. For the next seven decades, the park's ecosystem evolved without wolves, resulting in a huge increase in elk, which in turn had knock-on effects for the entire park. The elk overgrazed large tracts of land, resulting in a decline in forest land, and water flows, as well as the loss of habitats for species like songbirds and beavers. Then in 1995, following a heated debate among environmentalists, landowners, and park authorities, 14 wolves were reintroduced. Today there are about 100 wolves in the park, and we now know the extraordinary impact of their return. By controlling the elk population, against which only hunting by humans had provided a restraint, the wolves have reinstated the desired ecological balance. Reduced grazing pressure has allowed the forests to return, growing several times faster in only six years, resulting in more stable rivers (protecting river banks), and providing habitat for returning species such as eagles, badgers, and beavers. In just over a decade, the unique Yellowstone landscape was back in shape.

Previous pages: Mangrove forests in West Papua, Indonesia, are nurseries for fish and build resilience in tropical coastline ecosystems.

Another classic example of how sustainable management of ecosystems provides economic and social values to society comes from New York City. More than a century ago, officials had to make a decision: build a water-treatment plant to clean water flowing to the city or protect the 4,900 square km (1,900-square-mile) watershed, mostly in the Catskill Mountains of upstate New York, let the forest regulate the water flows, and use nature as the treatment service for the water supply. They chose the latter, and today the city's 8.4 million residents enjoy 4.5 billion m³ (1.2 billion gallons) of the nation's cleanest water every day. To replace the current system with a conventional treatment plant, consuming energy and chemicals, would cost as much as 10 billion USD, officials estimate. By using the forest as a water-regulator instead, the city saves many millions of dollars a year in reduced operational costs.

A more recent example of the value of ecosystem services came to light in 2012, when the UK sent scientists to Sweden on an unannounced mission to capture short-haired bumblebee queens. Once common in southern England, the bumblebees had vanished by 1988 after nearly all of the nation's wildflower meadows had been replaced by croplands. The irony of this development was that these same croplands depend on bumblebees and other insects for pollination, a service estimated to contribute at least £400 million a year to the UK economy. So the loss of bumblebees was a disaster.

A minor controversy erupted when the Swedish public discovered what the British scientists were up to, particularly as they had no official permission to collect Swedish bees. Sounding the alarm, local environmentalists warned that if the scientists collected too many Swedish bumblebees "we could end up in the same situation as the UK." One retired biologist fumed that the interlopers were "no longer the world's rulers as they were before when they just went around and took stuff." But as it turned out, in order to avoid a diplomatic skirmish between the UK and Sweden, the British scientists were quietly and rapidly given permission for their originally clandestine mission by Swedish authorities, and the controversy subsided. The British were then able to focus on the challenging job of reintroducing the bees to the English countryside, where conservationists had replanted areas with wildflowers, clover, and vetch. Nearby farmers also did their part, creating "green corridors" for the bumblebees along the margins of their fields, which they left undisturbed. They, more than anyone, wanted to see the bumblebees settle in back home.

The bumblebee incident was a reminder of the often unsung role played by so many species in landscapes and marine systems—and the cost to our economies when they're removed. As we mentioned in Chapter 3, the world is losing species today at a devastating pace, 100 to 1,000 times faster than the natural background rate. Unless this mass extinction is halted, the price tag could be too high to calculate.

CSR IS DEAD

More and more companies have reached the same conclusion: Sustainable business is good business. When General Electric announced that it has generated more than 160 billion USD in revenue, realizing 300 million USD in savings since 2005 by integrating energy efficiency into their production chain, it turned some heads. When they pointed out that, just as companies like Puma, Walmart, and Unilever are also doing, they make an increasingly significant portion of their net profit from sustainable business solutions, ranging from wind farms to solar voltaics and hyper-efficient turbines, it sent a clear message. The environment is no longer the domain of social or ethical responsibility in a company. It's increasingly becoming core business, the true heartbeat of the enterprise, the key to either dominating a market or disappearing.

A few visionary business leaders have known this for some time. For them, the triple bottom line of people, planet, and profit has always been an integral goal. But it has only been during the past three to five years that a major mindshift is emerging in the business world at large, as climate and ecosystems issues have moved out of the Corporate Social Responsibility (CSR) departments and into the boardrooms.

Speaking on a panel at the Stockholm Food Forum in May 2014, Peter Bakker, President of the World Business Council for Sustainable Development (WBCSD), a global network of 200 multinational companies representing about 10 percent of the world economy, declared that "CSR is dead." To survive in a world of rising competition for finite and increasingly dwindling resources, Bakker explained, of increasingly volatile and uncertain fossil fuel prices, of potentially destabilizing feedbacks from Earth for entire societies, we can no longer consider the planet as an external issue for businesses. Instead the planet is the business of business. Bakker's own organization has taken this to heart, transforming their so-called *Vision 2050* initiative into an Action 2020 plan that translates the science of planetary boundaries—including climate, biodiversity, water, land and nutrients—into science-based definitions of what green business means for the next few decades and beyond.

The EU recently finalized work within the so-called European Resource Efficiency Platform (EREP), which analyses the future competiveness of European industry. It has become increasingly clear, putting all the planetary risk cards on the table, that the only way for Europe to compete, creating growth and jobs in the future, is through major improvements in resources efficiency in the short term, and, in the long term, through a transformation to a circular economy. In a globalized world, where all citizens have a right to resources for development, 100-percent-green development just makes the most sense from the perspective of competiveness and the ethical responsibility of sharing. But the argument becomes even more persuasive if you consider the rapidly rising risks of global environmental disasters. Sustainability is the shortest path to prosperity, and it hinges on us being wise stewards of the remaining beauty in our ecosystems and planet as a whole. As Lord Nicolas Stern wisely pointed out at the World Economic Forum in 2014, sustainability is not a growth story, it is the only growth story for the world.

A NEW NARRATIVE

It's high time we changed the narrative of why we should care for our planet. It's long overdue, in fact, by at least 40 years. As "environmentalists" we're probably a big part of the problem. We've created a whole movement based on "protecting" the environment, and it has been so successful, it has contaminated everyone's thinking. We now live in a world where nature is placed on one side and society on the other. Environment versus development. And the two never meet. Economists are stuck in the obsolete mantra of addressing impacts on the planet as "externalities." Can you think of anything less accurate? How can you stand on a planet—one that is the source of all wealth—and declare it to be an externality?

In our attempts to solve environmental problems through civil, societal, business, and policy efforts, we get locked into the logic of "protecting" the environment from human damage. In UN climate negotiations, we talk about "protecting" the climate system and "burden sharing" to solve issues of responsibility. In the biodiversity convention, we focus on maximizing what's left to be conserved, to be "protected" from humans, appealing first and foremost to our ethical responsibility to protect other species. Businesses over the decades have responded very wisely to this state of affairs by creating a whole league of senior CSR officers and heads of the environment. Any company with global ambition has declared an altruistic engagement in "protecting" some element of the externalities affected by human action.

Well, this era is over. The story has changed. In the saturated and turbulent world of the Anthropocene, where we need to become stewards of the entire planet. Becoming planetary stewards means recognizing that our grand challenge is not about saving a species or an ecosystem. It's about saving us. It's about making it possible for humanity to continue pursuing economic development, prosperity, and good lives. The planet won't care if everything changes. It's our world that's at stake. Ultimately, any business must recognize that there can be no business in societies destabilized by abrupt social–ecological change. Only a stable climate and ecosystem can provide the resilience and sustainability we need to make our cities and villages livable.

Can we make the shift quickly enough? We think the answer is yes. Roughly 60 percent of the urban areas the world will need by 2030 have yet to be built. We know how to do that now in cost-effective ways that are smart from a climate, water, energy, and nutrient-recycling perspective. We also know how to plan urban areas for resilience with natural buffer zones to safeguard us against storms and floods. We know how to integrate ecosystems into dense urban areas to improve quality of life and diversity in ecological functions. The world will invest an estimated 90 trillion USD in new infrastructure over the coming decades. A mere four percent increase in that investment could make the entire infrastructure green from a climate perspective.

The only thing holding us back, in the end, is the obsolete but remarkably stubborn belief that what worked for us yesterday will work well tomorrow. We need a new paradigm of human prosperity within the safe operating space on Earth, growth within planetary boundaries, a development paradigm in which we succeed by becoming stewards of the remaining beauty on Earth, not as an aside, but as an integral part of our lives and businesses. We need to make it as natural as breathing. Once we do that, success in creating the base for thriving future generations will be that much closer.

A boy from the island of Batanta dives for shells in the waters off the Raja Ampat Islands, West Papua, Indonesia.

6

UNLEASHING INNOVATION

SPEAKING AT A CONGRESSIONAL HEARING in Washington, DC, in the 1950s, Kenneth Boulding made the provocative statement that "anyone who believes exponential growth can go on forever in a finite world is either a madman or an economist." Boulding, who was himself an economist, later developed the concept of "Spaceship Earth" to illustrate his alternative notion that humanity resides on a single and finite planet (an idea we see as an early precursor to our own planetary boundaries framework). Boulding characterized the limitless growth model as the "cowboy economy," since it evokes reckless, exploitative, romantic, and violent behavior. "In the cowboy economy, consumption is regarded as a good thing and production likewise," he told an environmental conference in 1966, "and the success of the cowboy economy is measured by the amount of the throughput..." By contrast, Boulding called the finite-world model the "spaceman economy," since "the earth has become a single spaceship, without unlimited reservoirs of anything, either for extraction or for pollution..." The goal of the spaceman economy wasn't greater levels of production and consumption, he said, but rather a higher quality of life.

Today, decades later, the debate about whether it's possible to have infinite growth in the face of rising environmental problems and resource constraints, continues to be a heated one. Our view is that neither neo-Malthusians like Boulding, who ridiculed the notion of growth on a finite planet, nor neo-liberal proponents of limitless growth, are right. We believe there's a middle ground between these two extremes. Based on the evidence before us, we're convinced that humanity's future lies in growth *within a safe operating space on Earth*.

For 40 years, like Moses in the unforgiving wilderness of the Sinai desert, environmentalists have wandered around advocating "limits to growth" with very little or no success. Now it's time to leave the wilderness and engage with the world in a much more constructive paradigm of "growth within limits."

Amphibious houses in Maasbommel, the Netherlands, are built with floating foundations that enable them to rise by up to 4 m (13 ft) in response to changes in the water level.

It all comes down to what we mean by growth. Lately, the word has been so mis-used, we'd prefer not to use it all, but rather focus on the ultimate goal of society, namely human wellbeing. In this sense, growth, or greater economic develop-ment, is but one means of achieving wellbeing.

For the vast majority of people on Earth who are just scratching by, there's a linear relationship between growth and human wellbeing. For this majority of citizens in the world—those between absolute poverty and middle-class (gener-ally making below 25,000 USD per year) greater wellbeing means improved life expectancy, education levels, access to health care, or social security. For the wealthy minority—those living mostly in the "old industrialized" nations (essentially the OECD nations) but increasingly also a rich minority in poorer nations—the relationship between conventional GDP-based economic growth and human development indicators is no longer straightforward. In fact, as Ste-fano Bartolino at Sienna University and others have shown, beyond a certain point of economic growth, individuals start losing social capital, which results in a vicious spiral as they compensate for this loss through even greater consump-tion, which leads to a rapid decline in natural capital.

In general, we agree that social and economic development is a necessary and desired means to reach humanity's goal of improved wellbeing. The big question is whether we can do right by both the people and the planet: Can we satisfy the aspirations of everyone in the world and still stay within a safe operating space of the planet? Right now, nobody has the answer. What we do know is that humani-ty's future prosperity hinges on safeguarding a stable planet.

A NEW WAVE OF INNOVATION

In their recent book *Abundance: The Future is Better than You Think*, Peter Dia-mandis and Steven Kotler argue that for the first time in history our capabilities have begun to catch up with our ambitions. Humanity, they say, is now entering a period of "radical transformation" in which novel technologies will enable us to meet the needs and aspiration of everyone in the world within a single gener-ation. Abundance for all, in other words, is actually within reach.

This would indeed be an astonishing step for humanity. As we stated earlier in this book, we too believe that we're living at a critical moment when the world's poor majority has a chance to become the rich majority—in a world with not seven but more than nine billion people by 2050. But to make this happen, the global economy is expected to grow three-fold during the same period. This will inevitably pose enormous challenges for a world already operating with a systemic 25 percent overshoot compared to its basic bio-capacity, a world in which humanity has already transgressed three planetary boundaries and Earth has started to respond with risky feedbacks and costly invoices.

Can it be done? Can we overcome the doldrums of inertia among world leaders in swinging the global development path toward sustainability? Are there enough promising advancements in exponential technologies, resource efficiencies, and circular economic models, as well as new preferences among young people for non-consumption-laden forms of happiness to make it possible? We're beginning to think the answer is yes.

The reason we're confident that systemic step-changes in efficiencies and solutions are now possible is the remarkable pace of recent technological breakthroughs. "Moore's Law" of computational progress—the observation in 1965 by Gordon E. Moore, co-founder of the Intel Corporation, that processing power for computers doubles every two years—continues to apply not only to microprocessors but also to advancements in biotechnology, nanotechnology, communications, and novel materials use such as graphene solar cells. The pace of science and technology, combined with rapidly spreading market-based societies, has inspired world leaders to set ambitious goals for development in the coming decades. For the first time, despite projected population increases of two–three billion people in Asia and Africa, including one billion absolute poor and three billion undernourished, our current generation of leaders believe it will be possible to eradicate absolute poverty and hunger in the world by 2030.

Not that technology can do it all. Rather, the challenges we face from negative global environmental impacts, especially the loss of biodiversity and climate change, are too great. A transition to sustainability can only be attained by combining technology with deep system innovations and lifestyle changes. But technology can play an important role, perhaps even a dominant role, in five key global transformations:

- *Renewable and sustainable energy systems*
- *Sustainable and healthy food systems*
- *Circular economic models for business, societies, and communities*
- *Sustainable urban futures in a world where 70 percent of all people live in cities*
- *Sustainable transportation systems*

These five areas of change are the prerequisites for sustainable stewardship of all the ecosystems on Earth, required to ensure future world prosperity. And we see evidence today that they can all be achieved through new technologies, novel system integration, and new behavioral strategies.

DECARBONIZING THE WORLD'S ENERGY SYSTEMS

On a typical Saturday morning in Germany, as you turn on your coffee machine, you're likely to get, on average, 30 percent—and under favorable conditions up to 75 percent—of your electricity from solar and wind power. This is an extraordinary development in the world's fourth-largest economy, made possible by a combination of innovative and long-term policy measures and advancements in the efficiency and cost-effectiveness of solar cell and wind power technology. Germany has combined in its *Energiewende*, the national energy plan adopted in 2011, a feed-in tariff system with an open market for buying and selling energy, strong political goals, and incentive-based measures that favor sustainable energy and punish unsustainable energy. The tariffs provide guaranteed minimum prices to utilities for energy supply, creating an incentive to invest in renewable technologies, while the decentralized energy market has added millions of small producers. Germany's political goals include phasing out nuclear power and achieving 80 percent renewable energy by 2050.

Of course there have been hiccups in the German experiment of "going to scale" with an energy system capable of operating within a safe climate space. The lack of an effective carbon price, for one thing, has created perverse rebound effects when the market gets flooded with cheap natural gas. Suddenly coal, the dirtiest of dirty fuels, becomes dead cheap. This clearly shows that technology and innovation cannot be left alone in a major transition. Policy must follow. The only way to succeed in a global energy transformation is to combine technological breakthroughs with science-based goals, such as the pace of decarbonization, and policies that encourage sustainability and punish unsustainability (such as a carbon tax, and legally binding emission targets).

There's ample evidence that the world can go to scale with a wide mix of sustainable renewable energy systems and innovations—and do so very rapidly. Solar cell technology, for example, has expanded exponentially in volume, efficiency, and cost-effectiveness. Although solar energy today accounts for less than one percent of the world's energy production, there's no reason to believe its exponential growth won't continue. Other promising breakthroughs include the use of graphene (carbon) as conductors instead of expensive and heavy silicon-based solar cells, and so-called Grätzel cells, a cheap dye-based solar cell technology.

We're not the only ones talking about a future in which all basic energy needs are met while operating within a global 2°C climate boundary. The Global Energy Assessment (GEA), the world's largest assessment of the world's energy future, also recently concluded that it can be achieved with today's technology. What it will take is a diverse palette of sustainable energy measures, ranging from major improvements in energy efficiency to biomass, wave energy, wind, solar, and various degrees of nuclear.

FOOD FOR ALL THROUGH SUSTAINABLE INTENSIFICATION

The way we produce food is the single largest "culprit" in our transgression of planetary boundaries. Food production consumes the most land and freshwater, emits the most greenhouse gases, represents the biggest threat to biodiversity, and is a key source of nutrient loading. If we can get it right on food, then we stand a very good chance of pursuing wellbeing within a safe operating space.

There's ample evidence that we can feed the world on our current existing cropland. This is of critical importance, since we now need to safeguard the ecological resilience of the remaining natural ecosystems on Earth. As we see it, this is the biggest incentive for innovation, since it will require a new doubly green revolution on existing farmland. To feed a world population of nine–ten billion by 2050, we'll need to increase food production by 50–70 percent—nothing less than a new green revolution. But now it must truly be green, occurring on existing farmland with no expansion, and this will require major breakthroughs in technologies and farming systems. We must radically enhance yields while also being sustainable, making crops even more resilient to extreme weather conditions such as droughts, floods, and dry spells, which are becoming more frequent. In an assessment led by Jonathan Foley of the University of Minnesota, together with a leading group of scientists, we recently concluded that it would be possible to feed humanity on current agricultural land by closing the yield gap between what can be produced and what is actually produced. This suggests a huge untapped potential for places like Africa, where yields tend to be on the order of 1 ton per hectare for staple food crops like maize and millet, when they could be 3–5 tons per hectare. We concluded that to succeed we need major farming innovations, such as precision farming and conservation agriculture, improved diets, and reduced waste, since we waste an amazing 25–30 percent of all food. So, it can be done. The question is whether it can be done in a sustainable way with regards to climate, water, chemicals, and nutrients, and whether we can build resilient farming systems that can deal with shocks from disease or weather.

Modern biotechnology can play a critical role in this regard. Combining genes into attractive combinations of healthy, resilient, sustainable, and productive food crops and species, without spreading to wild species or domestic equivalents, could be an essential part of the solution. Here we see the emergence of interesting potentials, moving away from the first generation of genetically modified organisms (GMO) associated with company dependence and

A vision of future prosperity is being built in Yaoundé, Cameroon, one of Africa's politically and socially stable nations.

uncertain side-effects, to step-changes toward sustainable and productive foods, from crops to fish. Transparent, public-sector science combined with tough government regulation is playing a key role in supporting modern biotechnology; for example, the development of perennial cereal crops that can evolve in harmony with natural ecosystems while providing nutritious food to rapidly growing populations.

TRANSFORMING TO A CIRCULAR ECONOMY

Our linear business models, in which we exploit resources and ecosystems, transform them into consumer goods and services, and then dispose of them, are obsolete. More often than not, this process involves leaking of energy and environmental flows along the value chain, causing various degrees of pollution and degradation. Instead we see a need to build a circular economy in which all capital (including social, natural, financial, and manufactured) involved in creating different products are recirculated into the economy and rebuilt for future generations of goods and services. As Ernst von Weizsäcker and his colleagues argue in their 2009 book, *Factor 5: Transforming the Global Economy through 80 Percent Improvements in the Global Economy*, resource efficiency could be improved five-fold if the majority of resources and ecosystem components in production are recirculated. This would take the world a long way toward a new logic for economic development—one that promotes growth without transgressing planetary boundaries.

A classic example was Ray Anderson's vision of a circular business strategy for his carpet tile company, Interface. Inspired by the Natural Step approach developed by Karl-Henrik Robert and his team over the previous 25 years, Anderson transformed his business with zero-emissions production systems, a radically reduced water and ecological footprint, and a high degree of recycling—leading to higher profits. His strategy was not to sell carpets but rather to sell a relationship with customers, maintaining responsibility for his products throughout their lifespan. Many other examples exist and will evolve in the future. The Ellen McArthur Foundation has recently presented ample evidence of the benefits for businesses to make the transition to circular business models. They point out, for example, that the remanufacturing costs of cell phones could be reduced by 50 percent if the industry produced phones easier to take apart, improved the reverse production cycle, and offered incentives to return phones.

BUILDING RESILIENT CITIES

We'll soon live in a world in which 70 percent of all people inhabit cities. Making urban environments livable, sustainable, and resilient to shocks and stresses, therefore, is one of the world's grand challenges. Fortunately, we know how to do this, taking advantage of dramatic improvements in energy and material efficiencies, and new approaches to urban landscape design that focus on resilience and quality of life.

"Well-designed cities can sustainably accommodate large numbers of people in a relatively small amount of space, offering improved quality of life and allowing for greater resource efficiency and the preservation of larger intact natural areas," wrote Secretary General Ban Ki-moon in a foreword to the 2012 *UN Cities Biodiversity Outlook*. Contrary to popular opinion, cities like Chicago, Mexico City, Singapore, and Cape Town actually support rich biodiversity. Urban centers are also sites of creativity, innovation, and learning. "Fostering these attributes is essential if the global challenge of preserving biodiversity in the face of unprecedented urbanization is to be met," wrote the authors, led by Thomas Elmqvist, one of our colleagues at the Stockholm Resilience Centre.

Investing in sustainable and resilient cities is one of the best ways of building future wealth and enabling prosperity within planetary boundaries. While the opportunities are large, the challenges are many times larger. The fastest-growing mega-cities and urban areas are located in some of the world's poorest regions, where effective governance is lacking, and where urban growth occurs uncontrolled and in a pace vastly outweighing the urban carrying capacity. But other cities are showing that change is possible through green infrastructure development, major investments in public transport, efficient recycling of waste, encouragement of renewable energies, promotion of car sharing, and optimization of natural sources of light. As examples, the UN study cited urban eco-areas such as Copenhagen's Vesterbro district, London's Beddington Zero Energy Development project, the Vauban neighborhood in Germany's Freiburg im Breisgau, and the Eva Lanxmeer quarter in the Dutch city of Culemborg. These areas are designed to be carbon neutral, the study explained, and "to promote concepts of eco-citizenship, encouraging people to improve their own wellbeing by preserving the environment."

SUSTAINABLE TRANSPORTATION

We now know how to build sustainable transportation systems by investing in large-scale networks of cycling highways and public transport. Cities like Copenhagen, Amsterdam, and Portland, Oregon, have been bike friendly for years. Lately they've been joined by cities in the emerging South. Bogota, Colombia, has greatly benefitted from the insightful vision of its Mayor Enrique Peñalosa who back in 1998 started a major effort to increase public transportation and cycling in the city. Today 300 km (186 mi) of bicycle lanes connect peri-urban areas, including slums, to the city center, increasing bicycle use five-fold since the initiative started. In New Delhi, where half of the city's residents regularly use public transportation, buses were converted between 2001 and 2003 to run on natural gas, reducing emissions of CO_2 and other pollutants. Bangkok's Skytrain, an elevated mass transit system that opened in 1999, now carries about 600,000 passengers a day. Future extensions are planned for the system, one of which is expected to open in 2017. Commuters in Seoul, South Korea, can use the same transit card in buses or taxis, making their daily journeys a little smoother. In Singapore, government officials have also invested heavily in promoting public transportation over private vehicles.

UNLEASHING CREATIVITY

Harvesting the benefits of the Second Machine Age—the name Erik Brynjolfsson and Andrew McAfee of MIT have given to the current era of technological innovation—will require a new paradigm for managing innovation and change. The idea of "managing" innovation may sound contradictory, since mechanisms of steering change, such as regulations, are often seen as hampering innovation. But we're arguing the reverse: We strongly believe that setting planetary boundaries for innovation and growth will unleash even more creativity, rather than stifle it.

The fact is, every disruptive new technology in the history of modern industry has resulted in rebound effects which, despite short-term improvements in efficiency, have led to negative impacts on the planet. We all know the classic examples. Refrigerator technology in the 1980s went through several Moore's Law-like leaps in energy efficiency, resulting in air conditioners and refrigerators that were not only less expensive but also massively more energy efficient. A win-win for people and planet. Or so we thought. But the net effect of this improvement has been a 15 percent rise in the size of fridges, and production of refrigerators has increased more than three-fold since 1995. As a result, what we won in efficiency we lost in absolute energy use.

Unfortunately, this rebound effect applies to essentially everything—even the darling of sustainable technology, communications. The mobile and digital revolution has ushered us into a new world, a transformative shift from the days when a wealthy minority communicated on monolog line-wired systems to today, when people everywhere use some four billion cell phones (including one billion smartphones with processing power greater than the supercomputers of the early 1980s). This revolution means we can all live more sustainably, by reducing travel and improving efficiency in communication. But we're now expected to change our cell phones every other year, at least, in a world where the number of phones has risen by a factor of 1.4 between 2012 and 2013.

What this tells us is that we need to embrace the technological revolution we're now experiencing, while also domesticating it. *Exponential technologies must have absolute bounds within which they're allowed to operate. Abundance needs to take place within the safe operating space of planetary boundaries.*

Will this hamper innovation? We think not. We believe it will actually unleash it. Generally, however, this is considered a controversial position. For many, particularly liberal economists, regulations of any kind are perceived to dampen entrepreneurship and as a consequence stifle business development and economic growth. We don't think this is the case for areas clearly associated with damaging the environment—and thus reducing our chances for long-term prosperity. Empirical evidence shows that tough regulation on environmental issues can speed up innovation and even reduce costs. Consider the example of catalytic converters in cars, or modern Freon-free cooling systems, or biomass-based heating technology and refrigerating. All of these technologies have made important step-changes under pressure from tough environmental regulation.

We're convinced that the best way to spur technological abundance in the areas of energy, food, health, and urban development, is to set long-term aspirational political goals, combined with regulatory frameworks and incentive-based policy measures that create boundary conditions for innovation challenges that meet sustainability requirements. It boils down to competition—to creating the equivalent of the XPRIZE challenge for the first commercial space travel—to unleash innovative technologies for sustainable management of the planet, where we not only make quantum leaps in terms of health improvements, services, and lifestyles, but also quantum leaps in terms of staying within finite biophysical budgets of our planetary boundaries.

POLICY MEASURES

What will it take to create this new paradigm—one in which innovation, technologies, collaboration, and universal values combine to promote a world community co-evolving with a stable and resilient planet? We envision a comprehensive set of policy measures in the following areas:

Men fish for stray golf balls near a golf course at the Hanoi Club Hotel in Hanoi, Vietnam; it's more profitable than fishing.

1. Global regulation of a safe operating space—based on defining science-based global sustainability criteria for world development (such as zero loss of biodiversity and staying under a 2°C danger limit for global warming).

2. Global agreements on an equitable sharing of the remaining biophysical space on Earth—including sharing responsibility for a finite global carbon budget, land budget, nitrogen and phosphorus budget, freshwater budget, and agreeing on safeguarding remaining critical forest systems and halting loss of biodiversity.

3. Putting a global price on carbon—at least 50 euros per ton (60 USD per ton) of CO_2.

4. Allowing for a wide diversity of policy measures and governance pathways—with "bottom up" alliances, pledges, citizen movements and activism, matching and evolving together with "top down" governance and institutional integration in nations and regions around the world.

5. Going "beyond GDP" to define new criteria for growth and progress—building on concepts of green economic development along with new metrics to measure progress.

6. Investing massively in capacity development—including open transfer of technologies to the world's developing nations and major investment funds to take the Second Machine Age to scale.

We think the time has finally come when we can take such measures from theory to practice, not at a small scale, but at a grand scale. And we're not alone. A growing number of business leaders, policymakers, and citizens in general have reached the same conclusion that the world is now facing unacceptable risks, and that we need to find ways to promote development within global sustainability criteria. Many top CEOs of large multinational companies, for example, are now willing to accept a price on carbon, as long as it is global and predictable. Similarly, many policymakers are willing to scrap GDP as the only measure of economic progress, and follow nations like China in the pursuit of alternative metrics, such as gross ecosystem product (GEP) which will operate in parallel to GDP and measure the increase and decline in stocks and flows of natural capital. These are just a few examples of the significant progress we've made in the last four to five years in accounting for global environmental risks. The conversation has gradually moved from being about burden-sharing and contraction—about protecting and regulating—to strategies for minimizing risk, generating benefits, and developing modern high-technology solutions that generate human prosperity within the safe operating space of a stable and resilient planet.

SECTION 3

SUSTAINABLE SOLUTIONS

IT WAS POSSIBLY the toughest moment in my professional life. In late July 2009 a meeting on the environment was about to begin in Åre, a beautiful ski resort 650 km (400 mi) north of Stockholm. It was a gathering not only of environment ministers but also of industry ministers and commissioners. José Barroso, president of the European Commission, was coming as well.

At the time, I headed the Stockholm Environment Institute (SEI), Sweden's most influential climate and environmental policy research institute. Although the institute is an independent research organization, I was eager to support the Swedish government in its leadership on climate issues. So, together with Jacqueline McGlade, head of the European Environment Agency (EEA) in Copenhagen, I wrote an opinion piece for Sweden's largest daily newspaper, *Dagens Nyheter*, for publication before the Åre meeting.

In the piece, McGlade and I argued that political leadership on climate issues must be based on science. Politicians have a tendency to compromise with science, and climate change has been no exception. As many studies have shown, including our planetary boundaries analysis, the world needs to keep atmospheric concentrations of carbon dioxide (CO_2) from rising above 350 ppm to avoid severe impacts from global warming. Some politicians have argued that a CO_2 concentration as high as 450 ppm would be safe, under the assumption that it would lead to an increase of only 2°C (3.6°F) in the average global temperature, and that the consequences might not be so bad. There is, in fact, very little scientific evidence to link 450 ppm with a 2°C increase or to say that such an increase would be safe. On the contrary, there's plenty of evidence that a 2°C increase would be costly and dangerous, and might even trigger catastrophic tipping points. We pointed this out in our piece.

Our big mistake was to quote José Barroso. A few months earlier, he'd defended the 450 ppm figure by saying that's "what science tells us." We simply pointed out that, in fact, there was a large degree of scientific uncertainty about that number and that, if we want to be reasonably safe, we should aim lower. Well, our piece appeared just before the Åre meeting was to begin. But it was interrupted by a very angry Barroso, who asked what on Earth was being said in this Swedish newspaper article signed by "his" director of the EEA. He demanded that the meeting wait until the article was translated into English.

When Barroso understood the article's full content, he became furious. Meanwhile, McGlade called me to ask why we hadn't gone over the Barroso quote in detail. The article had been approved by her media folks, but I had to admit that I hadn't asked her specifically about the quote. The Åre meeting was held up until McGlade publicly announced that she'd been manipulated into signing the article. (Later, she wrote a letter to Barroso, explaining that, although she denounced the quote, the article was scientifically correct.)

The incident shook me up at the time. But now I realize what an important reminder it was that scientists must stay true to the science and not compromise with it or make it politically palatable. Although I'm still sorry about the fuss it created, I have no regrets about what we wrote in the newspaper. Our role is to communicate the facts, not to favor what is perceived to be politically possible. Using science as an excuse for inadequate action is never acceptable.

I think the reason why we see a growing discrepancy between what science states is necessary and what politics claims is possible is that politicians are afraid that ambitious sustainability goals will threaten economic growth. There's very little to support this fear. On the contrary, environmental policies such as taxes actually stimulate innovation and new growth. The Swedish carbon tax, for example—the highest in the world, with a price of approximately 100 euros per ton of CO_2—hasn't destroyed the Swedish economy. On the contrary, it has stimulated both economic growth and green technology innovation.

As we'll see in the following chapters, a grand challenge for humanity is to find ways of unleashing the power of nature-based solutions and innovation within the safe operating space of science-based goals. We may need new forms of governance at the planetary level to achieve such goals. We'll most certainly need new instruments to measure our progress in strengthening Earth's resilience. But we're confident that the key to meeting such goals will be to turn them into incentives for new, exponential technologies—brilliant, cool, and profitable ideas to open the pathway to a sustainable common future.

7

RETHINKING STEWARDSHIP

WHAT DOES IT MEAN these days to talk about environmental stewardship? How can we safeguard the planet if the daily actions of billions of people are undermining Earth's basic functions?

Ask Moustapha Amadou, an elderly farmer in southwestern Niger, where the life-giving summer rains now arrive later, sometimes several weeks later, than they used to. Amadou lives on the edge of the Sahel in the small village of Samadey, about 100 km (62 mi) north of Niamey, the capital of Niger. During the past 20 years, farmers in this region have become increasingly anxious about what they've been witnessing—even more variability in an already variable rainfall regime. Something's changing, they realize, and it's happening quickly.

After the first heavy downpour in June, the savannah around Samadey smells of warm, sweet tropical fruit. The gray, dusty expanse has turned into a muddy brown landscape. Children are playing in the flooded lowland at the center of the village, recalibrating their bodies and minds from the dry to the wet season. There's an aura of celebration in the community, a sense of being blessed.

Before he does anything else, Amadou checks to see if the rain has moistened the soil in his field to a depth of at least one hand's length. Then, and only then, following a practice that goes back more generations than can be remembered, will he decide if the time has finally come to plant the precious millet seeds he's kept stored away during the long, hungry dry season.

Things don't seem quite right to Amadou. For one thing, the lowland shouldn't be so drenched after the first rainfall. The land surrounding the village should absorb at least the first four to five rainfalls, filling up the 62-m (203-ft) deep, hand-dug, village well. The villagers rely on that well for precious freshwater throughout the eight-month dry season. But for some time now, the rains have fallen with a higher-than-normal intensity. Instead of soaking into the soil, as every farmer wants them to do, the rains from these downpours wash over

What does the future hold for a boy in Rwanda? Sustainable solutions offer him and the rest of his generation the best chance to alleviate poverty.

land, triggering erosion, which every farmer dreads. This is the new curse.

The gradual warming of Earth has not only shifted rainfall patterns, it has also increased the likelihood of sudden downpours, floods, heat waves, droughts, and wildfires. Under such circumstances, it's not enough any more for small farmers like Amadou to manage their local environment. He, like all of us, must learn how to become planetary stewards.

But what does that mean exactly? And how do we achieve it? The answer, we believe, involves a rapid transition from a way of life based on exploiting natural resources for our own benefit to one based on strengthening Earth's resilience. Whether we're farmers like Amadou, factory workers in Shanghai, fishermen in Indonesia, or shopkeepers in Des Moines, Iowa, we all depend on functioning lakes, forests, waterfalls, oceans, and glaciers for our wellbeing and prosperity. No matter how modernized we think we are, how alienated from the natural world, none of us can get by without thriving ecosystems all around us.

Farmers like Amadou are a step ahead of most of us urban dwellers, because his livelihood is so directly rooted in the variable swings of local weather. The challenge for the rest of us is to recognize that we also depend on sustainable management of watersheds, river basins, and large biomes such as glaciers and forests, as well as the entire climate system. The rainfall in Samadey is intricately linked to the way we manage ecosystems and the climate across the planet.

NEW RULES OF THE GAME

The challenge we all face now—to pursue a prosperous future for everyone within the safe operating space of planetary boundaries—calls for bold new strategies for governance at both the global and local levels. It won't be enough to apply the "top down" power of institutions, enforcement agencies, global justice systems, international partnerships, new trade rules, or global regulations to demand changes at the planetary level. Nor will it be enough for "bottom up" grassroots activists, community managers, business innovators, educational experimenters, or public–private organizers to work their magic at the local level. We're going to need both approaches. And we need them to interlink and work together.

Fortunately, these two strategies can support each other, with "top down" forces creatively stimulating "bottom up" ones, and vice versa. Clear rules from the top, for example, repeatedly asked for by business leaders, can create clarity at the bottom, stimulating investments in sustainable and resilient business strategies. We just need to establish the new rules of the game to get things started.

It's become obvious, for one thing, that some form of strengthened governance at the planetary scale will be required to achieve the kind of transformative changes the world needs. In a world where the majority of inhabitants have yet to claim their fair right to development, strengthened global governance will be necessary to secure a just distribution of ecological space. We're going to need global guardians of planetary boundaries—not as a means to "rule the world," impose a cap on development, or limit growth, but rather to make sure we don't derail Earth from its current stable condition.

Right now, we have nothing of the sort. In fact, leaders across the world remain skeptical about the need for governance of the Earth system. But the game has changed. We no longer have the luxury of debating whether or not we need to make ends meet at the global scale. We can't afford to argue any more about whether or not to stay within a common global carbon budget, for example, or a land budget, or a freshwater budget. Knowing what we know now about the risks of crossing catastrophic tipping points, we must strengthen our capacity to meet global sustainability targets.

Despite the good intentions and tremendous efforts of men and women around the world who have signed more than 900 environmental treaties during the past 40 years, we're still moving in the wrong direction. We still lack any mechanism for planetary-scale governance with the teeth to guide the development of our modern economy. The "tragedy of the commons"—in which shared resources are squandered by selfish but, given the rules of the game, "rational" behaviors—has never been more pronounced than today, when separate nations pursuing self-interests have kept the global curves of negative environmental change all pointing the wrong way.

The problem is that the old approach no longer works. In the Anthropocene, when humanity has saturated Earth's capacity to buffer our pressures, one nation's unsustainable growth is another nation's threat to growth. We shovel environmental risks across the world in a way that wasn't the case a decade or two ago. Global challenges require global solutions. Development at the local, national, and regional scales must now be capped by globally agreed-upon sustainability boundaries and targets. That's why it's become necessary to strengthen global governance on environment and development, which will support local actions and innovations.

But strengthening governance at the top doesn't mean weakening it at the bottom. On the contrary, we're convinced that we need to transform societies both at the bottom and at the top, creating a dual pressure on beneficial processes leading to a sustainable world. One approach might be to strengthen the multilateral governance system within the UN. The United Nations Environment Program (UNEP), for example, should be transformed into a specialized agency with regulatory mandates at the global scale (like the WTO and WHO)—a United Nations Environment Organization

(UNEO)—with strategies to empower local communities and businesses.

This upgrading has been discussed over the years, and in 2012 UNEP took a small step in that direction by establishing a universal membership in its governing council and providing a more stable funding base. We must now be brave and experiment with systemic shifts in the way we govern for global sustainability. The aim would be to universally agree on global sustainability criteria, such as planetary boundaries, that are monitored, reported, and enforced, with the objective of fostering not only a step-change in innovation, experimentation, learning, and practices to support human prosperity, but also a just way of sharing ecological space with all fellow citizens in the world, while staying within safe planetary boundaries.

It's not for us to stipulate what these practices and innovations should be. But we do see an urgent need for a clearly defined playing field where all nations, businesses, and communities know the rules of the game. We believe a new set of global regulatory measures will be necessary, ranging from a global carbon tax to international agreements on all planetary boundaries. We don't believe these will hamper the operations of market economies or stifle innovations and technology leaps. On the contrary, we all know that markets are social constructs. They've always required a "helping hand" to keep them focused on their primary role, to provide human wellbeing.

New legal regulations, norms, and values will be needed, including international recognition of every individual's right to a resilient and well-functioning Earth system. Effective compliance regimes will also be required to compel collective action. We also urgently require a new global strategy for sustainability, building upon the voluntary action plan known as Agenda 21. Adopted at the UN Conference on Environment and Development (UNCED) held in Rio de Janeiro, Brazil, in 1992, Agenda 21 offered ideas to local, state, and national governments for sustainable ways to combat poverty and pollution while conserving natural resources.

A key step in this direction is to support current efforts to transform the UN Millennium Development Goals (MDGs) into globally agreed-upon Sustainable Development Goals (SDGs). The eight MDGs, established as a result of the UN Millennium Summit in 2000, have been remarkably successful during the past decade in stimulating and focusing international efforts to alleviate poverty and hunger, among other important goals. Now we're turning a corner. We're starting to see the contours of a new paradigm building on the concept of development within the safe operating space of a stable planet.

It started with the Ban Ki-moon high-level panel on global sustainability, which formed the basis for the UN Earth summit in 2012 (the UN Rio+20

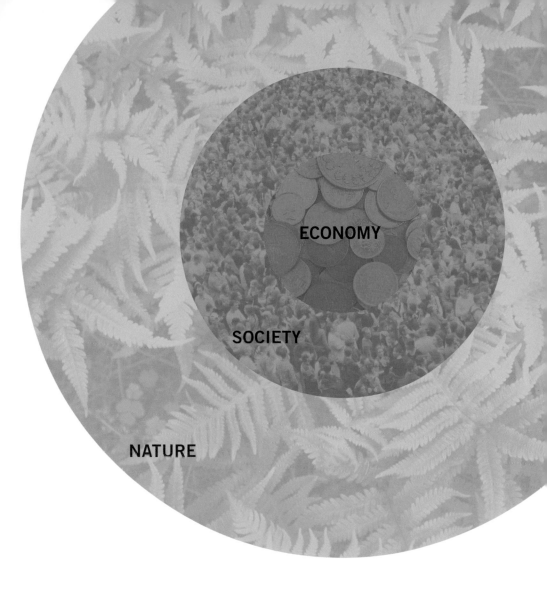

Figure 7.1 Shifting to a New Paradigm for Sustainable Development in the Anthropocene.
Attaining world-development aspirations—from eradicating poverty and hunger to economic
growth—requires that the world evolves within the safe and just operating space of a resilient
and stable Earth system: abundance within planetary boundaries. This changes our paradigm for
development, away from the current sectoral-pillar approach of social, economic, and ecological
development, as separate parts often seen as contradictory forces, with the economy advancing at
the expense of natural and social capital. Now we must transition toward a world logic where the
economy serves society so that it evolves within the safe operating space on Earth.

conference). For the first time, world leaders discussed the fact that global sustainability is now a prerequisite for poverty alleviation and that rising global environmental risks may undermine future progress in development. The UN Open Working Group (OWG), advancing the draft plan for the SDGs, has carried these insights forward in 2013–2014, and the final set of proposed SDGs actually includes the contours of a development agenda within planetary boundaries.

Most important, the proposed SDGs include an aspirational set of goals that constitute what we could call the social "end game" for humanity. The objective is not to reduce poverty and hunger by 50 percent, but to eradicate them entirely. The new agenda is about enabling education, health services, gender equity, and transparent governance for all. The 17 proposed goals—and more than 150 targets—include global sustainability goals that essentially cover all nine planetary boundaries. There are explicit goals on oceans, climate, ecosystems, and freshwater. Taken together, the SDGs show that we're entering a new development logic, based on a new narrative, in which world leaders recognize that human progress hinges on Earth resilience.

Such an integrated SDG framework requires that we abandon our old approach to sustainable development, based on the three separate pillars of social, economic, and ecological sustainability. Instead, as we recently showed in a scientific paper produced by a team led by David Griggs of the Australia National University, we must adopt a nested development framework in which the economy is a method to serve society, which in turn develops within the safe operating space of planetary boundaries.

A NEW SHADE OF GREEN

At the same time that policymakers have been struggling to negotiate new approaches to global governance (with meager results so far), more and more business leaders have realized that "business-as-usual" is no longer an option. Until recently, too many executives operated under the assumption that "environmental stewardship" meant wrapping their profit-focused business core in a flowery shield of green consciousness. This superficial camouflage resulted in various attempts to "green-wash" annual reports and results. But today we see a shift. The Anthropocene has entered the boardrooms, not as a threat but as strategic "insider" information about the risks and opportunities of tomorrow's markets—in simple terms, as core business.

Consider, for example, the cornerstone report, *Vision 2050*, by the World Business Council for Sustainable Development (WBCSD), which has recently been translated into an Action 2020 strategy. While the report laid out a pathway for business to reach sustainability by 2050, the Action 2020 document is a result of a profound effort to translate the science on planetary boundaries into actionable targets for business. Specifically, Action 2020 identifies a set of "must-have" immediate actions to enable businesses to progress within a safe operating space, by staying within a science-based carbon budget, raising agricultural outputs on existing land without increasing water-use and nutrient-loading, halting deforestation, and biodiversity loss.

Similarly, the "Shell Energy Scenarios to 2050" describes an era of "revolutionary transitions," including a global resource crunch, global environmental risks, and momentous growth in energy demand. The latest scenarios offered a "blueprint" for a sustainable energy future that meets both social and ecological requirements. In the same vein, "Getting into the Right Lane for 2050," by the National Environmental Assessment Agency of the Netherlands (PBL) and the Stockholm Resilience Centre, assesses the feasibility of implementing a sustainable vision for Europe. It also shows that Europe can get into the right lane—within a safe operating space with respect to climate, energy, land, and biodiversity—while supporting economic development.

Probably the most comprehensive vision for future global sustainability, however, is a scenario analysis by the Great Transitions Initiative (GTI) led by the Tellus Institute. This analysis, covering most of the planetary boundaries, shows how challenging it is to provide a fair future for all within the world's safe operating space. Even with the most ambitious and optimistic implementation of all the policy options within our current development paradigm, we show that humanity will face great difficulties in staying within safe levels of all planetary boundaries.

To us, this conclusion demonstrates two things: First, that action now is possible; and second, that "tweaking and tuning" of our current social and economic systems won't be enough. We're going to need immediate actions that aim at far-reaching systemic changes of our economies and institutions, our values and lifestyles, supported by a broad mind-shift that can trigger such a great transformation. The critical thing to remember is that these transformative changes have only one aim: to improve our prospects for growth and prosperity in the future, not to hamper them. Remember, without Earth resilience, there will be no world prosperity. Can transformational change happen fast enough? Well, in truth we do not know, but there are inspiring examples of social tipping points toward sustainable wealth.

One such example at the national level was what happened in Australia's Great Barrier Reef Marine Park a few decades ago. Like many people around the world, Australians were struggling to manage vulnerable coral reef ecosystems

Walking together with nature, a mother and child stroll along Leblon Beach in Rio de Janeiro, Brazil.

in the face of climate change, polluted runoff, overfishing, and other human and natural pressures. Confronted with declines in the populations of dugongs, turtles, sharks, and other fish, it became clear to everyone who knew the reef that the original management system was no longer adequate.

One of the most controversial initiatives proposed at the time was to extend the zone closed to all forms of fishing from 6 to 33 percent of the total reef area— creating the largest no-take zone in the world. A critical step in the process was to convince local communities that the reef was facing many threats, and to enlist public support for protecting a larger area of the reef and managing it more flexibly. This was accomplished through a major "Reef Under Pressure" community consultation campaign, supported by a steady flow of science from leading international coral-reef scientists like Professor Terry Hughes of the ARC Centre of Excellence for Coral Reef Studies at James Cook University in Australia, and co-author of our original planetary boundaries research.

The successful campaign to rezone the Great Barrier Reef Marine Park has been recognized as a groundbreaking international model for better resilience management of the oceans. In a 2008 study, researchers from the Stockholm Resilience Centre and the ARC Centre of Excellence for Coral Reef Studies concluded that the campaign represented a "mental tipping point" for the public, in which the earlier perception of a pristine reef shifted into a growing awareness that the ecosystem was approaching a critical point of no return. This shift in thinking led to a more integrated view of humans and nature, based on active and flexible stewardship of marine ecosystems for human wellbeing. The bottom line, according to our science peers, was that laws alone can't bring about the changes necessary to protect the world's ecosystems. Good science and public support are also vital.

MEASURE, MEASURE, MEASURE

What we don't understand, we ignore. What we don't measure, we don't manage. For the past 30 years or so, ignoring environmental risks has worked out quite well for us. We saw few if any signs of the social–environmental turbulence now playing out at the global scale. But those days are over. It's time to open our eyes. We must both measure and understand what's happening to Earth.

We believe this to be a key step in the mind-shift required to reconnect with the biosphere. We must measure every aspect of how nature interacts with societies and track changes in the stability of the Earth system in real time. The architecture for this effort already exists in many places. Scientists and agencies around the world have established a Global Earth Observation System of Systems (GEOSS) covering many of the key planetary boundary processes, from changes in the climate to the chemical composition of the oceans. Earth-observing satellites monitor changes in sea level, and a global mesh of floating Argo buoys report continuously on shifts in temperature and acidity in our oceans, among other things.

We understand much better now how Earth operates. Yet there are still frustratingly large gaps in our knowledge. Data on the pace and global distribution of sea level rise, for example, remains incomplete and uncertain. So does information about the energy exchange between oceans and atmosphere in the global ocean conveyor belt; the global rate of biodiversity loss; actual melt of ice sheets in Antarctica, the Arctic, and Greenland; cloud dynamics and local shifts in rainfall as a result of global changes in weather patterns; and the importance of air pollution for the global climate. To put it simply, we need to measure, measure, measure, not to get certainty, but to understand risk and become cleverer stewards of nature for our own benefit.

A key priority is to track biodiversity loss. Why? Because biodiversity plays such a critical role in the resilience of ecosystems. There are still major gaps in our knowledge about species richness in the world. We're losing massive amounts of biodiversity without ever knowing what we are losing.

But measuring alone won't be enough. We must also share with the public our knowledge of how Earth works and changes, showing how it affects and changes our societies. This will require a deeper integration between natural sciences and social sciences. Luckily there is a growing momentum around the world at high-quality research institutions to address the interconnected links between humans and nature. And great things are happening here, such as the launch of Future Earth, the world's largest international science endeavor on Earth system research for global sustainability. Future Earth aims to foster new knowledge to support a transformation to global sustainability with stakeholders outside of science.

We also need to connect the dots between observations, understanding, and society. Here we have a big gap. We still live in a world with islands of insight in an ocean of ignorance on both the risks and opportunities in a sustainable world. We believe the public's lack of awareness about the global risks we face is largely due to the shortage of user-friendly, attractive, and widely accessible information, and that new scientific insights advance so fast that educational systems can't keep up. That makes education our number one priority. If every secondary school classroom in the world were to tear down its geological charts and replace them with one that includes the Anthropocene, we would already have come a long way. We need new curricula for high schools and undergraduate programs at universities. Above all, we need to train a new generation of economists who understand that generating wellbeing for society means keeping the economy within planetary boundaries. As Kate Raworth at Oxford University has pointed out, if every student studying economics would start his or her program learning about planetary boundaries, the world would look different.

The environmental agenda deserves to be elevated from its "second-rate" position in politics and business. This is already gradually happening. In the global risk reports in recent years by the World Economic Forum, leading executives put global environmental risks near the top of the things they worry about, both in terms of their likelihood of occurring and in terms of the seriousness of the consequences for their businesses. Among the threats they identified were water supply crises, food shortages, extreme volatility in energy and agriculture prices, rising greenhouse gas emissions, and failure to adapt to climate change. Other threats mentioned were terrorism, chronic fiscal imbalances, and severe income disparities. This tells us something about the rapidly changing risk landscape. As these executives realize, we're living in an increasingly turbulent world where social and ecological crises are not only equally likely, they're also interconnected, potentially interacting and reinforcing one another.

When environmental risks are properly understood, that is, they become security risks, stability risks, foreign policy risks, financial and business risks. Developing a strategy to manage environmental problems becomes a question of risk management. Compare that to what we see in global markets today. The only risks markets appear to be interested in, officially at least, are economic ones. The possibility that businesses could lose market shares and profit margins is taken so seriously that companies on the stock market provide quarterly reports on their performance.

To us, this seems shortsighted. By fixating on short-term profit, businesses ignore long-term growth of social–ecological welfare.

For that reason, we think it would be a good idea to phase out quarterly financial reports, and replace them with bi-annual ones, or replace the current system with quarterly reports on how well businesses are doing in meeting targets aimed at staying within safe planetary boundaries. These should include accounting for carbon, nitrogen, phosphorus, water use, and other impacts on ecosystem services. (Advanced tools already exist for this, such as the Corporate Ecosystem Services Review, developed by the World Resources Institute.) A natural evolution of such a business reconnection to the biosphere would be to see "state of the planet" pages next to the business pages in our daily newspapers and news programs. Imagine if there were Earth pages next to the financial pages in the *Wall Street Journal*, presenting quarterly reports on greenhouse gas emissions and use of ecosystem services among the largest companies in the world.

Finally, we believe grand experiments are needed to share knowledge with

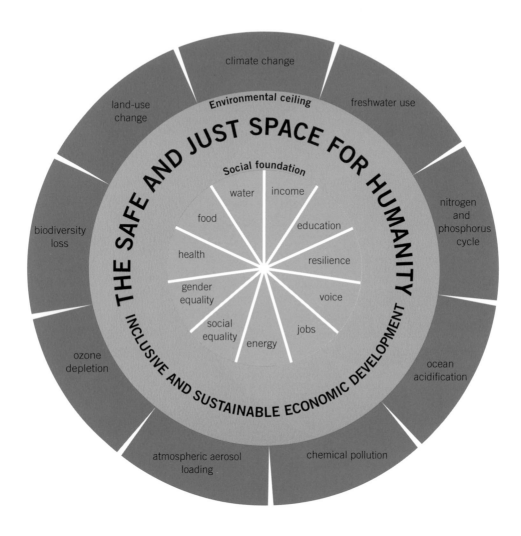

Figure 7.2 A Safe and Just Space for Humanity. If there is a biophysical ceiling for human development on Earth—the planetary boundaries—then there must be a social floor, as pointed out by Kate Raworth, then at Oxfam, now at Oxford University. The social floor should consist of the universal human rights of access to resources, ecosystems, atmospheric space, a stable climate, and the dimensions of equity, dignity, resilience, and agency that are associated with a good life.

society at large. One idea would be to develop a global network of Earth Situation Rooms—physical or virtual—where citizens across the world could follow, in real time, what's happening to the planet. Picture a wide circular room, surrounded by large video screens connected to the world's most advanced Earth-observing satellites and monitoring systems. Here, in one place, citizens could catch up, in real time, on the environmental state of the planet, as well as regional trends, with respect to all of the planetary boundaries and key biomes.

GOODBYE GLOBAL COMMONS

It might seem paradoxical in this epoch of escalating threats to the planet that we should put such a strong emphasis on preserving the beauty of nature at the local scale, or that we should lobby on behalf of the diversity of landscapes, or that we should urge local actions by individuals, communities, and other sectors. But it all starts where we stand. As Mahatma Gandhi is said to have advised, you must "be the change you want to see in the world." That means we must build from the "bottom up", whether we're trying to save the jaguar in Latin America, the wolf in the Nordic countries, or the tiger in South Asia; whether we're trying to secure freshwater supplies for a village in India, or plant a field with millet seeds outside a small village in Niger.

It no longer makes sense to us, in fact, to keep talking about a "global commons" or about economic "externalities." These terms for the complex environment "out there" were developed for the old paradigm, where humans and nature were seen as separate entities. Today, in our big world on a small planet, we're all part of the same commons, and changes in our shared environment —such as the climate system, the ozone layer, or the global hydrological cycle—rebound directly on local economies. In today's interconnected and environmentally saturated world, there are no externalities. Everything, from finite resources to clean air, forms an integral part of our efforts to generate human wellbeing.

That's why we say there aren't any global commons any more. Because of the feedbacks generated by the Earth system as we push environmental processes too far, every part of our surroundings has become a personal responsibility for each one of us. A global commons is something that no one owns, and therefore is poorly managed. In the Anthropocene we all "own" every bit of the global commons and have a responsibility for them—because their fate will determine our future.

Following pages: The Milky Way photographed from Cameroon.

8

A DUAL-TRACK STRATEGY

HERE'S SOME GOOD NEWS: There's a mountain of evidence that a sustainable future is actually possible. Many technologies, practices, and systems already exist to bring about necessary global transformations. The Anthropocene, after all, doesn't have to be an epoch of negative human impacts only. It could be a "good" Anthropocene, in which massive human innovation guides the world into an era of abundance within planetary boundaries. This story, of health and prosperity for all within the confines of a stable planet, has never been told before. The time has come to do so.

We believe it starts with a mind-shift, a change in perspective. As we saw in Section 2, the old ways of thinking have to go. The idea that economic growth is disconnected from nature, for example, or that environmental issues stand in the way of human development, are obsolete notions. In fact, just the opposite is true: To provide a sustainable basis for human prosperity, we must reconnect our societies with the biosphere and strengthen its resilience. The systems that support life on Earth, from a stable climate to rich biodiversity, are prerequisites for modern economies. In the Anthropocene, sustainability is the key to prosperity.

But such a mind-shift could take time, perhaps even generations, to achieve. And that's a luxury we don't have. To stay on the safe side of most planetary boundaries, we must reduce the rate of negative environmental impacts within this decade. We can't wait 30 years. Unless we do something right now to lessen the risks of triggering self-accelerating Earth processes, we're likely to suffer catastrophic outcomes within this century. For that reason, we're proposing that, as political leaders, business managers, and private citizens, we take a dual-track strategy: 1) tackle the most urgent challenges with immediate fixes, while 2) doing everything we can to promote the long-term mind-shift that this book is all about: How to reconnect our human

Plantations in Sarawak have replaced diverse, resilient rainforest with monoculture palm oil trees, creating short-term benefits but uncertain returns when subject to shocks.

societies—and what we value in life—with the beauty of nature and the resilience of Earth.

Let's address the first of these goals, the short-term fixes. We see two obvious opportunities for humanity to jump-start the future immediately. One is a global transformation to a world economy free of fossil fuels and the other is a global transformation to a sustainable food future. Both would solve many of the grand challenges facing humanity, generate social and economic benefits, create stability and security, and yield positive synergies across many planetary boundaries. In both cases, we already have the know-how to succeed. To unleash investments, innovation, and policies that enable these immediate fast-track transformations to occur the world community must agree to meet our need for modern energy and food while staying within planetary boundaries. That means defending the 2°C limit for global warming that world leaders agreed upon in 2009 in Copenhagen (although science tells us 1.5°C would be safer), which gives us a global carbon budget within which we must operate. We must also agree to produce our food within planetary boundaries, where the goals are relatively straightforward: zero loss of biodiversity and zero expansion of agricultural land, while keeping rivers flowing, and closing the loop on nitrogen and phosphorus.

POWER UP

The first fast-track solution is a switch to renewable energy systems. Based on what we know today, this should be doable by mid-century. For one thing, the sustainable potential of renewable energy is enormous, many times more than what we need to power the whole world. The current global primary energy consumption amounts to approximately 500 exajoules (EJ), and the sustainable potential for wind power alone exceeds 1,000 EJ. If you add biomass, solar, geothermal, and hydropower, the sustainable potential of all renewables exceeds 11,000 EJ. Additional capacity could be claimed through changes in technologies or practices that boost energy efficiency, many of which would also save money.

A recent study by the German Advisory Council on Global Change (WBGU) painted a positive picture of our energy future. It showed that a total phase-out of crude oil, coal, and natural gas could be achieved by 2050 through technically feasible strategies while still meeting growing global energy needs. According to the report, this transformation could be accomplished if humanity were to stop using fossil fuels for electric power generation, heating, and transportation systems (the toughest challenge, no doubt). Heating and cooling would be achieved by using electric heat pumps, solar thermal energy, and combined heat and power technology (CHP), in which waste heat is recovered from power plants or factories for secondary uses. Vehicles powered by electricity, hydrogen, methane, natural gas, or fuel cells would replace those using gasoline. The plan also calls for efficiency improvements and savings in consumption, with a one percent reduction per year in global demand for heating and cooling, and a one percent increase per year in demand for electric power. It won't be easy, in other words, but it can be done.

The Global Energy Assessment (GEA), a major international study of the world's energy future, in 2012 presented similar results. The study showed that it would be possible to decarbonize the world's energy systems by 2050 while meeting rising energy demand, although it will require major investments in renewable energy systems. The GEA estimated that some 1,500 billion USD per yr would be enough to spur massive investments in clean-energy technology. The current subsidies for fossil energy amounts to an astounding 500–600 billion USD per yr, or about a third of the investment needed to unleash a global transformation to a clean-energy future within a climate boundary. So just phasing out fossil-fuel subsidies, a promise already made by the G20 countries, would go a long way toward opening the window for a renewable energy future. Moreover, these are small numbers compared to the overall global GDP. As shown by the fifth assessment of the IPCC Working Group III, mitigation efforts to stay within 2°C of global temperature increase would slow down economic growth by a mere 0.06 percent.

This was hammered home in late 2014 by the Global Commission on the Economy and Climate, led by Felipe Calderón, which convincingly showed in its "New Climate Economy" report that there is no contradiction between economic growth and a transition to a decarbonized world economy. On the contrary, most analyses today show that pathways to decarbonization involve profitable investments with low short-term costs and high returns in the medium and long term.

The best way to kick-start this global energy revolution, we believe, would be a global price on carbon. Most global energy analyses suggest that a price in the range of 50 to 100 USD per ton of CO_2 would be necessary by 2050. But in the European emissions trading scheme (ETS) the price has been hovering below 20 USD per ton of CO_2, which clearly is too low to stimulate transformative change. Only Sweden has operated a systematically high carbon price for a long period. Since 1990, Sweden has had a comprehensive carbon tax across all energy sectors of approximately 100 USD per ton of CO_2. This tax has resulted in a decoupling of the Swedish economy, allowing

continued economic growth while reducing national CO_2 emissions from industry. It has also resulted in a transformation of the heating sector, shedding off the last fossil energy sources in favor of heat production based on biomass residue from the forest industry. Placing the right price on carbon can thus trigger an energy transformation.

Why are we so confident the time has come to make this transition? One reason is that renewable energy technologies have now reached a market penetration big enough to enable them to take the big leap into dominating the energy market. Empirical evidence shows that new technologies require a market penetration of approximately 10 percent before they can accelerate to a dominant position. For decades, despite very rapid growth rates, renewable energy systems, from photovoltaics to wind turbines, have accounted for only single digit percentages of various energy markets. But this is rapidly changing. In Germany and several other countries, renewable energy systems have already approached the 10 percent threshold in market penetration.

Another reason, of course, is the urgency of reducing global CO_2 emissions. As we saw in chapter 2, for Earth to have a reasonable chance of staying below an additional 2°C of warming, the world's economies must decarbonize by 2050. After that, in the second half of this century, we must have negative emissions (by sequestering more carbon than we emit, for example, in land and biomass). Two thirds of the long-lived greenhouse gas emissions, or 78 percent of CO_2 emissions, come from the use of fossil fuels in our energy systems. If we can solve this problem, we can solve a large portion of the climate challenge.

A global energy transition, in other words, would solve the bulk of the global climate crisis, provide energy to poor nations as well as rich, and expand the many alternatives to fossil fuels we have today. Technically and economically, we're ready to jump-start a sustainable future.

A TRIPLY GREEN REVOLUTION

The second fast-track solution is a transition to sustainable agriculture. This should also be doable by mid-century, when a world population of some nine billion people will require 50 percent more food. Agriculture today represents the single largest cause of biodiversity loss and greenhouse gas emissions (about 30 percent of global GHG emissions originate from agricultural

Feeding the world through sustainable agriculture will require both bioscience and indigenous knowledge, as can be found here in the Dominican Republic.

production, roughly half from cultivation and the other half from deforestation). It's also the world's largest user of land (almost 40 percent of the world's terrestrial surface is under agriculture), and the largest user of freshwater (70 percent of withdrawals of freshwater from rivers are used for irrigation). In addition, agriculture is the main source of nutrient overload from leakage of nitrogen and phosphorus into our waterways. What we eat has a much larger cost than what we pay.

Agriculture's heavy footprint derives from its modernization since the 1950s. The "green revolution" that brought such a remarkable increase in productivity during the second half of the 20th century was based on fertilizers, fossil fuels for tractors and food processing, and an extensive use of chemicals. What we need now is a new kind of revolution—a "triply green" revolution that boosts productivity even higher, but also reduces impacts on the environment and sustainably manages water resources. This revolution will require a global partnership between science, farmers, businesses, and societies in which we reduce agriculture's dependence on fossil fuels (for traction, transport, processing, and fertilizer production), close the loop on nutrient flows (what goes out needs to go back in), and strengthen the resilience of the environment (for example through improved water productivity). Among the keys to achieving these goals will be reducing food waste (as much as 30 percent is lost between farm and fork), rethinking our diets (especially with respect to meat consumption), and improving water management (through some old-fashioned ways to collect rainfall and store runoff).

In many regions of the world, there's still a huge gap between current crop yields and what could be achieved through modern techniques. In large parts of Africa's savannah regions, for example, average yields of staple food crops such as maize, sorghum, and millet remain around 1–2 metric tons per hectare, even though it should be possible to generate yields on the order of 4–6 tons per hectare.

A major challenge is securing a reliable water supply. Although these regions generally have enough water in total, the bulk often comes in a few great downpours, with a high risk of droughts and floods with long dry spells in between. In these smallholder, rain-fed agricultural systems, less than 50 percent of the rainfall is generally used to produce food. Instead, it's lost through evaporation and runoff, which generally causes soil erosion and land degradation. Improved water management, ranging from soil and water

Previous pages: Every square meter of available land is cultivated in Rwanda, the most densely populated country in Africa.

conservation to small-scale irrigation systems, could go a long way toward enabling big improvements in productivity. Investing in strategies to manage rainfall—such as collecting runoff for supplementary irrigation in water-harvesting micro-dams that store excess runoff water during intensive rainfall events—can reduce risks enough to trigger investments in other improved practices.

But water measures alone won't do it. Farmers also need more nutrients. Compared to farmers in the USA or Europe, who apply more than 100 kg (220 pounds) of nitrogen and phosphorus to each hectare per year, many African farmers apply less than 10 kg (22 pounds) to each hectare per year, even though they lose more than 50 kg (110 pounds) per hectare of these chemicals when the harvest is removed. No wonder, then, that such soils eventually degrade and lose their productivity.

The "triply green" solution for these farmers isn't to dump more fertilizers on their fields, but rather to adopt sustainable and affordable practices for improving soil, nutrient, and water management. Take plowing, for example. Although it effectively reduces weeds, plowing also exposes the richest part of the soil, the top layer, to heat and erosion. In tropical regions, this contributes to rapid burning of organic matter, which not only releases CO_2 but also reduces the water-holding capacity of the soil, its rooting depth, and its capacity to absorb rainfall and avoid erosion. The effect of plowing is thus a gradual degradation of soil fertility.

By contrast, when farmers adopt conservation tillage the soil is not turned with a plow. Instead it's opened along planting lines to a depth of at least 15 cm (six inches), which exceeds the traditional plowing depth. This creates a "micro ditch" where rainwater is concentrated and enables farmers to carry out precision application of manure and fertilizers. The idea is to copy nature as much as possible to build up organic matter and biological activity, which in turn raises productivity.

No-tillage practices have caught on lately in places such as Ghana, Zimbabwe, Niger, Kenya, and Tanzania. About 25 percent of US farmers have also replaced plow-based techniques with zero-tillage methods, while more than 70 percent of farmers in Latin American countries such as Uruguay, Paraguay, and Bolivia have done the same. As Rattan Lal of Ohio State University has shown, adopting such practices could shift agriculture from being a major source of carbon emissions to becoming a carbon sink, sequestering potentially 1 Gt C/yr (gigaton, nine zeros), which is more than 10 percent of current global emissions.

Another way to enrich soils is to recycle wastes. Our societies today remain far too linear in terms of agriculture, with nitrogen and phosphorus fertilizers plugged in at the farm end, and waste and leaking nutrients coming out the other end, polluting our freshwater and marine systems. The challenge is to close nutrient loops. Productive sanitation systems provide one such strategy. Almost all of the nitrogen and phosphorus we eat returns to the environment through our excreta and waste. Using ecological sanitation systems that separate urine and feces, and different systems to recycle treated waste from urban areas to return nutrients back to agricultural land (where it came from), we can reduce nutrient loads, provide affordable fertilization, and reduce the pressure on finite phosphorus resources. Closing the loop on nutrient cycles is a necessary strategy to return us to a safe operating space with respect to nitrogen and phosphorus. At the same time, we should recognize that many poorer regions, where fertilizers are not being used enough, will need to significantly increase inputs of such nutrients.

Advancements in biotechnology will also play an important role in a triply green revolution. The scale and pace of the world food challenge is so great, we'll need new breakthroughs in genetic crop research to succeed. This could be accomplished, for example, through the development of perennial cereal crops to drastically reduce tillage requirements, make the systems resilient to droughts and dry spells (because of their deeper root systems), and immediately transform cropland from carbon sources to carbon sinks. Similarly, genetic strains with drought resistance, high nutritional value, or short growing periods may soon be combined into single food crops.

To sum up our short-term fixes, then, we believe that humanity can jumpstart a transition to a sustainable world by focusing on renewable energy and agricultural reform. But to bring about lasting change we must also inspire a long-term shift in perspective that ignites a new generation of leadership, renewal, and innovation.

THE BLUE MARBLE

As the *Apollo 8* spacecraft orbited the moon on Christmas Eve in 1968, Earth appeared to rise from below the horizon like a blue marble in the vast blackness of space. For the first time, humanity saw our home planet as a single fragile sphere. "The vast loneliness is awe-inspiring and it makes you realize just what you have back there on Earth," astronaut Jim Lovell said. The photograph astronaut William Anders took of the "Earthrise" (see page 3) became one of the most famous in history, suddenly bringing home to many millions around the world the importance of preserving our shared planetary inheritance.

Sadly, in recent decades we've forgotten that message. In the midst of an unprecedented growth in the world economy, we've largely ignored the rapid deterioration of Earth. We've forgotten that a stable climate, adequate freshwater supplies, clean air, and biodiversity are all generated by a functioning stratosphere, atmosphere, hydrosphere, biosphere, and cryosphere. Instead, we've benefited from mistreating the planet, perceiving environmental protection as an economic cost and thus a burden to growth. We've convinced ourselves that we must choose between growth or sustainability, but cannot have both.

As we saw in Chapter 3, the consequences of holding onto these invalid assumptions about the relationship between nature and society are now beginning to be felt. We're pushing, or at risk of pushing, the boundaries for climate change, freshwater use, land use, nutrient overloading, air pollution, biodiversity loss, and chemical pollution, well beyond the ranges that would safely maintain a stable Earth. We can't continue to operate the planet like a subprime loan, taking advantage of Earth to live beyond our means.

The time has come to put an accurate value on natural capital in economic terms. For some time now, we've failed to attribute the full cost of our production and consumption systems. In simple terms, we've been cheating ourselves. At the same time that we've increased GDP, which is essentially the only aggregate measure of economic progress that we have, we've also degraded land, polluted air, destroyed water-supply catchments, cut down rainforests, and contributed to the melting of polar ice sheets. We must reverse that relationship by recognizing that the global economy is, in fact, a subsystem of the biosphere. To serve humanity well, the economy needs to operate within the confines of Earth's life-support systems, not only for future generations but also for the stability and security of nations today.

As cracks have appeared in the current global development model, it's become increasingly clear that we need a better way to measure human progress, one that moves beyond GDP as an indicator of wellbeing. There's a growing recognition that global sustainability, equity, resilience, and happiness must all be important parts of a new definition of human development. The current metrics are insufficient for the task. We must begin to implement a broader social–ecological perspective on human wellbeing in richer nations (that have reached saturation levels in growth), while investing in efficient and effective growth to alleviate poverty in poorer nations.

Finding ways to foster human wellbeing on a crowded, increasingly wealthy planet is the newest challenge for humanity. In 1990, an estimated 42 percent of the world's population was living in absolute poverty (earning

Hydroelectric projects such as the Bakun Dam in Sarawak generate electricity with low
carbon emissions, but they also impact ethnic minorities and natural habitats.

less than 1.25 USD per person per day). By 2015, that number may be as low as 10 percent. Annual growth rates in double digits since 2005 have lifted out of poverty an estimated 430 million people in South Asia and 250 million in East Asia. But as we saw earlier, research increasingly shows that we cannot sustain our gains against global poverty without also safeguarding ecosystem services and attaining global sustainability. The progress we've made in combating poverty today may be undermined tomorrow by environmental feedbacks from Earth due to growing human pressures on the planet.

Under today's development paradigm, every nation fends for itself in a market where no one takes responsibility for Earth's shared ecological space, the global commons we argued in Chapter 7 no longer exists. Earth is largely seen as a free lunch, where the greedy grab the bulk of what they want, and the hungry stand by empty-handed. What's needed in the Anthropocene is a just, rights-based, and equitable sharing of the world's remaining ecological space.

Toward that end, Oxfam has recently developed an integrated paradigm, based on the planetary boundaries idea, to address both the social and biophysical challenges of safe and just human development. The ceiling for development would be set by the planetary boundaries (how much ecological space we can appropriate). The safe operating space under the ceiling would then be given a floor, corresponding to basic and universal human requirements for a good life. Meeting these requirements (for food, shelter, health, energy, education, resilience, and security) takes up a certain portion of the natural capital and the Earth system's services (for example, a portion of the remaining global carbon budget to stay under 350 ppm of CO_2, a certain amount of land and water for food, a certain amount of nitrogen and phosphorus, and so on). This social core would be a non-negotiable universal right to ecological space on Earth. The remaining part, between the biophysical ceiling and the social floor of the planetary boundaries, constitutes the degree of freedom, within the safe operating space, for meeting aspirations beyond basic needs.

This human development paradigm is simple, but drastically different from what we have today. Its goal is to ensure that we share social and natural capital in a just way without undermining the prosperity of fellow citizens or future generations.

A SHREWD INVESTMENT

One of the most important insights of the planetary boundaries framework is the recognition that we should stop thinking about certain things as costs or burdens for societies and see them for what they truly are: long-term natural venture capital for prosperity and wealth creation.

For two decades, delegates from around the world have argued over the most beneficial "burden-sharing" regime for the UN's framework convention on climate change. Beneficial for whom? For themselves, never for the world. The same goes for air pollution standards, critical loads for chemicals, restrictions on use of heavy metals, negotiations to curb global deforestation, or the endless efforts within the UN convention on biological diversity to come to grips with the global extinction of species—so far to no avail. Why? Because national leaders create strategies based on short-term cost-benefit analyses, in both political and economic terms. And Earth so far, almost always, gets undervalued.

This might be understandable if there were any evidence that sustainable management of natural resources, living ecosystems, and the global climate were indeed a burden to our economies and societies. But there isn't any such evidence. Nowhere. It's a myth we chose to cultivate. Even in conventional economics we know that production processes depreciate capital, that eroding assets is a cost, while protecting and maintaining assets is a long-term benefit. The same goes for Earth.

The evidence is overwhelming, dating back to 18th-century economists such as David Ricardo and Adam Smith. Back then, land was a key source of wealth. But as technological revolutions over the years gradually reduced the value of land in relation to other industrial capitals, this fundamental insight was lost. Today we know these economists had it right. Our common planetary capital—including a stable climate and ecosystem services, from provisioning of food from agricultural land to regulation of freshwater flows to mega-cities—should be the basis of the forms of capital in our economy.

A shift in our economic perspective is thus 200 years overdue. The key first step in making that shift, apart from putting an economic value on natural capital, would be to incorporate benefits from sustainable practices into the equation. The question shouldn't be: What's the cost of moving toward a low-carbon society. It should be: What different kinds of benefits will investments in low-carbon energy systems, transportation, and food production generate for families, sectors, nations, and regions?

Perceptive business leaders have also picked up on the fact that additional benefits are likely to follow from investing early in a transition to sustainable business models and sustainable nations. They recognize that the real risk for organizations could be getting left behind. We live, after all, in an increasingly turbulent world, where crossing tipping points and "peak everything"

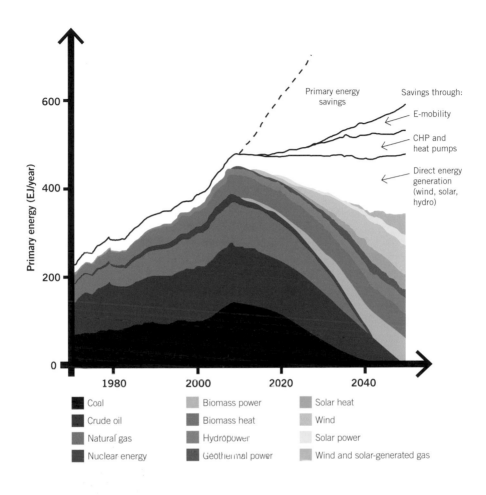

Figure 8.1 Exploring Avenues for a Decarbonized World. Many analyses today show, as illustrated here, that a world transformation to a renewable energy future is possible by the second half of this century. Returning to the safe space for climate will require action on many fronts, combining major expansions of renewable energy sources such as wind, solar, geothermal, hydro, and biomass, with major efficiency improvements and changes in behavior. Without a global energy transformation we will follow the dotted business-as-usual line. CHP is "combined heat and power," highly efficient processes that capture waste heat in conventional systems, such as steam clouds rising from cooling towers, and integrate it back into production of usable heat and power.

are imminent threats, resulting in price volatilities, unacceptable risks of disastrous outcomes, and a scramble for resources. The real cost for a business or a nation, therefore, may be to stubbornly stay put in the current dirty, unhealthy, inefficient, and increasingly unattractive, growth model. Nations and businesses that run ahead of the crowd, transitioning to closed-loop production systems and renewable energy sources, may well be the biggest winners of tomorrow.

Jobs are an integral part of a dual-track strategy for growth within planetary boundaries. In Ho Chi Minh City, Vietnam, a young cook stir-fries meat from a cobra.

9

SOLUTIONS FROM NATURE

LOOK AROUND YOU, wherever you are, whether it's in a car, an airplane, an office, or on a cozy sofa in your living room. Everything you see comes from nature. Every material, from rare earth metals to textiles, plastics to wood, is extracted from the biosphere. The food we eat, the way we cool or heat our homes, even our latest smartphones (with an average of more than 50 different metals) all come from services provided by nature. It should come as no surprise, then, that innovations and far-reaching solutions to today's problems are also increasingly emanating from nature.

Consider the humble thistle. In Sardinia, where the prickly weed has invaded abandoned wheat fields, biotech researchers have made a surprising discovery. Vegetable oil from the purple-flowered *Cynara cardunculus*, sometimes known as the artichoke thistle, can be converted into a wide variety of monomers and intermediates, which form the base products for numerous industries, from tires to lubricants and cosmetics. When one of the region's most polluting petrochemical plants shut down in 2011, the decision was made to transform it into one of the world's most advanced and innovative green chemistry and bio-refinery plants. Instead of using fossil fuels, the new refinery, which opened in Porto Torres in January 2014, uses wild and cultivated thistles to make products for the bioplastics industry.

This bold project came at a good time. Italy had just made a federal decision to ban fossil-fuel-based plastic bags, allowing only biodegradable bags, a move that could help turn the new weed-based venture into an exponential technology. Italians, it turns out, use more than 300 bags per person each year, which adds up to more than 20 billion bags. By phasing out plastic bags, Italy hopes to cut back the amount of plastic waste and fossil energy use among consumers, reducing long-term accumulation of micro-plastics in the oceans. This far-reaching regulation will certainly be an inspiring incentive for the advanced bioplastics industry.

Many such examples of nature-based innovations are now emerging, from

A old stone wall separates farmland from grazing land at Öland, Sweden, where sheep interact with diverse flora and fauna. A rapidly changing environment puts such ecosystems at risk.

the use of silkworms to weave fibers for high-strength ropes for sailboats to exploration of snake venom to treat heart disease. In addition, nature-based technologies are also being adapted to radically reduce environmental impacts, from biotechnological and genetic changes in plants to electric cars and passive houses with zero-energy consumption. Still another application of nature-based concepts is what companies are doing to adopt circular business models or reduce waste by recycling products like aluminum cans. Clearly we need major breakthroughs on at least three fronts—exponential technologies from nature, innovations to reduce impacts, and wider system changes—to navigate a sustainable future of growth within limits.

Many of these solutions are well established but haven't yet been taken to scale. Take wind and solar power, for example. Although both are now competitive economically with fossil-fuel-based supplies of electricity and heat, they still provide only two percent of global energy use. The good news is that we're seeing exponential growth of wind and solar in many parts of the world, and in countries such as Germany, with the world's fourth-largest economy, renewable energy provides about a fourth of the electricity (2013), of which 12–13 percent comes from solar and wind.

In the field of solar power, technological advancements during the past few decades have opened many new possibilities. In the 1990s the main challenge was the cost of solar panels. Today, given the rapid development of inexpensive silica-based technologies, that's no longer a problem. As Marika Edoff, an engineering professor at Uppsala University, says, the primary challenge now is storing energy in a cost-efficient way. For many years the technological development was driven by companies in Germany, Spain, and Italy, but today the advances are coming from China, Japan, and Brazil. In Africa, where the infrastructure often doesn't exist for electric power transmission, several countries have turned to solar panels to produce power where it is needed.

Less well known are some of the other untapped opportunities nature offers. In terms of generating abundance within planetary boundaries, these can be even more effective than the most ambitious circular economic models. As pointed out by the author Günter Pauli, who wrote about nature-based solutions in his book *Blue Economy*, if you were to improve an unsustainable system by 90 percent, it would still be 10 percent bad. But if you switched, for example, from a fossil-fuel-based material like plastic to a biological material like bamboo, you'd be going from an always more or less bad system to a

Saving the majestic Bengal tiger and its habitat is a smart strategy to secure long-term water-supply and other ecosystem services for humankind.

potentially 100 percent good system. Wellbeing gained. No growth lost. Such disruptive sustainable solutions can fundamentally transform our modern societies from being more or less unsustainable to being fully sustainable.

LEARNING TO LOVE MAGGOTS

Of the many nature-based technologies that Günter Pauli cites in *Blue Economy*, one of the most intriguing concerns the maligned insect larvae we call maggots. As our diets have become more animal intensive, food production has steadily increased the pressure humanity exerts on water, land, nutrients, biodiversity, and climate. In fact, our hunger for meat is one of the main reasons why world agriculture is threatening planetary boundaries. Our appetite for meat also creates a great deal of waste. Almost half of the animals we slaughter for consumption ends up as waste—in Europe about 150 kg (330 pounds) per person. But now innovative projects have developed maggot farms to decompose animal waste, in the process producing low-cost maggot protein which is then used as animal feed. In parallel, medical research indicates that maggots can be a very cost-effective method of removing dead tissue from wounds and potentially also stimulating cell growth. As an exponential sustainable technology, in other words, maggot farms could open up a whole new avenue of insect-based waste management and health treatment.

We're also learning to love snakes (if we didn't already). Among the most widely used groups of drugs in Western society today are those to lower blood pressure. Many work by inhibiting "angiotensin converting enzyme" (ACE), preventing it from constricting blood vessels and thus raising blood pressure. What many people don't know is that the most widely used ACE-inhibiting drugs, such as Captopril, were initially modeled on a component of venom from the Brazilian pit viper *Bothrops jararaca*. The idea came from observations of the viper hunting, when researchers discovered that prey fell to the ground due to a sudden drop in blood pressure. The naturally occurring component of the venom, a peptide called teprotide, wasn't suitable for use as a drug. But careful and creative studies made it possible to understand exactly what part of the peptide was active, eventually giving us the drug Captopril.

As an example of a major industry adopting nature-based business models, the garment industry stands out. The textile business for many years has striven to increase its share of sustainable fabrics, such as sourcing cotton from ecological farming systems and reducing pollution from the chemical dye industry. More and more large firms like H&M, the Swedish clothing company, have

Following pages: A Red Coral grouper swims by Red Sun Coral. Besides providing more than 250 million people with food, reef systems are essential for many otherwise pelagic species.

realized that this isn't enough. They understand that the long-term goal must be to close the loop on textile flows through the recycling of textile fibers. Piloting is ongoing among many companies to produce garments not only from recycled cotton and wool but also from other post-consumer waste materials such as bottles made of polyethylene terephthalate (PET) or other recycled plastics. Although the scale of such recycling is still small—only about 20 percent of fabrics can be recycled using existing technology—this is clearly an area where consumer values can spearhead rapid changes, as it becomes increasingly unacceptable to allow the bulk of old cloths to end up in landfills.

In another promising trend, companies such as Puma, the German-based sportswear and shoe manufacturer, are explicitly recognizing the degree to which their businesses depend on natural capital and ecosystem services. When Puma's chairman, Jochen Zeitz, launched his company's Environmental Profit and Loss (EP&L) program in 2011, he described it as a key strategy in sharing information with Puma customers. By putting price tags on everything from T-shirts (where 20 percent of the cost is a subsidy from nature) to sports shoes, showing the environmental cost of producing them, the company created a new platform for sustainable engagement with consumers.

As Zeitz later pointed out, the work also guided business strategy within Puma to identify the most sustainable and profitable investments, such as using recyclable materials for shoes. This work, together with initiatives by Richard Branson of the Virgin Group, including the Carbon War Room and establishing the Elders network, inspired these two business leaders to establish The B Team, which includes a dozen or so business leaders such as Paul Polman of Unilever and Ratan Tata of the Tata Group. The most interesting thing about The B Team is their belief that "global business leaders need to come together to advance the wellbeing of people and the planet." In fact, they say, business has to think in this way in order to thrive. This is in line with what many fields of science say, and is at the heart of the mind-shift we're focusing on in this book. Launching EP&L accounting and other similar initiatives are steps in the right direction.

CONNECTING VISION WITH SOLUTIONS

The Baltic Sea is sick. In fact, it may be the world's sickest inland sea. Because of an overload of nitrogen and phosphorus from agricultural runoff and decades of loading toxins from urban and industrial waste, the Baltic had crossed a tipping point by 1989, incidentally exactly at the point in time we have argued the world tipped over from being a small world on a big planet to a big world on a small planet, becoming locked in an unfortunate new ecological regime. From a cod-rich, nutrient-poor, and oxygen-rich state, it flipped over into the current murky

mess with a poor stock of large fish and an overdose of nutrients. Cyanobacteria, so-called blue-green algae, feed on the nutrients and grow exponentially, because most of the zooplankton that used to graze on the blue-green algae has been eaten by herring and sprat, whose numbers exploded after their predator, the cod, disappeared. Now climate change is making things even worse, by speeding up the melting of ice from the Arctic riparian basins in the north. This adds more freshwater to the Baltic's brackish marine environment and raises water temperatures—both processes that reinforce the sea's current sick state. About a sixth of the Baltic has actually become a "dead zone", the largest such area on the planet, with very low levels of available oxygen.

Shockingly, this environmental disaster has occurred right in front of the citizens, governments and businesses of the nine riparian states—Sweden, Finland, Denmark, Estonia, Latvia, Lithuania, Russia, Poland and Germany—that generally value the Baltic highly. In fact, recent studies by the Baltic Stern project, an international research program, have shown a high willingness among citizens of these nations to pay for a healthy Baltic. And the benefits of doing so would be considerable. A recent analysis by the Boston Consulting Group (BCG) shows that a sustainable scenario of investments for the Baltic Sea could generate 550,000 jobs and 32 billion euros in economic value added by 2030.

What would it take to turn this situation around? It boils down to operating within planetary boundaries, including a radical reduction of nutrient loading from agriculture and urban effluent. As an important first step, St. Petersburg, the single largest source of pollution into the Baltic Sea, opened its first modern wastewater treatment plant in late 2013. Finally, a new fishing management regime is needed to safeguard diversity and restore top predators like cod and pike—a regime supported by both inhabitants and fishermen, just as marine national parks have in Kosterfjorden or Kenting. The key, as BCG pointed out, is that all the stakeholders in the nine nations sharing the Baltic must agree on a sustainable vision. This is the mind-shift needed to guide a transformation: An agreement that the beauty and resilience of thriving Baltic ecosystems are the basis for human wellbeing and economic development, and that policy, business, and citizens all gain from investing in this shared vision.

Another transformation in sustainability is taking place in the world's urban areas, where many of the most important investments in nature-based solutions are likely to take place. The reason is simple. We have crossed a tipping point, with more than half of the world's population living in cities. There are now 28 mega-cities with more than ten million inhabitants, a number expected to rise to 40 cities by 2030. By 2050 almost two thirds of everyone on Earth will live in cities, which means 2.5 billion new urban inhabitants.

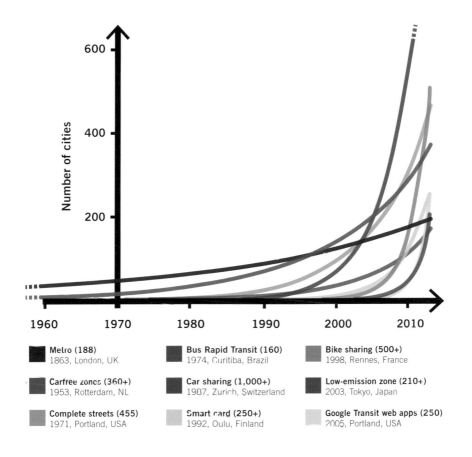

- ■ Metro (188)
  1863, London, UK
- ■ Carfree zones (360+)
  1953, Rotterdam, NL
- ■ Complete streets (455)
  1971, Portland, USA
- ■ Bus Rapid Transit (160)
  1974, Curitiba, Brazil
- ■ Car sharing (1,000+)
  1987, Zurich, Switzerland
- ■ Smart card (250+)
  1992, Oulu, Finland
- ■ Bike sharing (500+)
  1998, Rennes, France
- ■ Low-emission zone (210+)
  2003, Tokyo, Japan
- ■ Google Transit web apps (250)
  2005, Portland, USA

Figure 9.1 Smart Transportation Systems. Since 2000 a range of innovative and sustainable transportation systems have been launched in cities around the world, from rapid bus transport to car- and bike-sharing. Many of these systems take advantage of existing technologies. Bus Rapid Transit (BRT), for example, uses dedicated lanes, pre-boarding ticketing, and custom-designed bus stations to turn an old technology into an effective new form of mass transit. First pioneered in Curitiba, Brazil, BRT is now operating in more than 166 cities worldwide.

A mountain of evidence, including the recent UN Cities Biodiversity Outlook, now suggests that urban ecosystems with a relatively rich biodiversity can help to buffer cities from extreme events such as Hurricane Sandy, which hit New York, or the kind of disastrous landslides that have hit communities in Taiwan. And, in our globalized world, businesses depend on sustainable cities to thrive. As Mats Lorentzon, CEO of the music-streaming company Spotify, pointed out in an informal dialogue recently, his organization employs young talent from all corners of the planet, and a key factor that attracts these individuals to work in their Stockholm offices is the city's clean and safe environment. In other words, an area's sustainability—its resilience, health, and beauty—has become one of its most important characteristics, and that means that cities need to operate within planetary boundaries to thrive in the future.

Just look at Singapore, one of the world's most densely populated urban environments, which has proven that compact living can be surrounded by ecosystems that provide both recreation and resilience. What always strikes us when visiting Singapore is how much nature is embedded within this mega-dense city environment. The growth of such cities, we would argue, is a social tipping point that we're crossing right now. And the trend is positive, with the number rising exponentially of urban areas adopting strategies aimed at more livable environments.

By contrast, the 20 million people living in São Paulo, Brazil's largest megacity, have recently faced major environmental problems. The worst drought in 80 years has led to unprecedented water scarcity in the city. Some municipal reservoirs have even run dry. In the words of Vincente Andreu, president of Brazil's water regulatory agency, residents should brace for "a collapse like we've never seen before" if the drought continues. What's causing the crisis? According to Brazilian scientists such as Carlos Nobre, a leading authority on climate change, the decline in rainfall in the region is most likely linked to global warming and deforestation of the Amazon rainforest. Amazonia is a massive water vapor pump, releasing an estimated 20 billion tons into the atmosphere each day from vegetation, a large part of which falls back on the forest as rainfall. A significant portion of this moisture moves downwind to supply rainfall to the catchments that fill São Paulo's water reservoirs with drinking water and water for irrigation. Brazil's National Space Research Institute (INPE) concluded recently that the humidity from the Amazon, in the form of "flying rivers," has dropped dramatically, contributing to the current drought.

It is thus increasingly clear that São Paulo, as Brazil's financial and business heart, can only beat successfully if it receives life-giving rainfall from a sustainably managed Amazonia. At the moment, things don't look encouraging. Deforestation in the Amazon region jumped by an astounding 29 percent between 2012 and 2013, the first increase since 2008. Studies have shown that continued deforestation could lead to a 20 percent reduction in dry season rainfall on average by mid-century, and that we can't rule out the possibility that the rainforest is about to cross a tipping point, shifting irreversibly into a drier savannah system.

This would not only undermine cities like São Paulo and the Brazilian economy, it would also ultimately affect the entire world, because the planet would lose one of its grand carbon sinks and water vapor pumps. This represents yet another reason why we must halt global deforestation, which studies have already shown would have a large and immediate benefit for humanity in terms of climate mitigation, reducing emissions by some 5 billion tons of carbon per year, compared with the annual emissions of 32 billion tons from the burning of fossil fuels.

Taking care of the rainforest, in short, provides an obvious win–win by both securing immediate economic growth in cities like São Paulo and providing insurance against future catastrophic climate risks—all based on working with nature, preserving biodiversity, and building long-term resilience.

MAKING IT WORK

As we've shown several times in this book, working with nature is a key strategy for developing sustainable systems that operate within planetary boundaries. This is especially true of agriculture. The key objective is to transform farming systems from a source to a sink of carbon, because in so doing, soils enhance their nutrient- and water-holding capacities and become more productive and less prone to land degradation. We know how to do this with numerous strategies, from adopting conservation agriculture systems to closing loops on nutrients through balancing livestock with crop cultivation, crop rotations, and smarter precision farming and overall integrated nutrient management. Even in the harshest of environments we see success.

In Niger, one of the world's poorest nations, nature-based solutions have improved the livelihoods of more than a million households. Despite the fact that people here live in one of the least productive and most water-scarce savannahs on the planet, farmers in the Maradi and Zinder regions in southern Niger have increased agricultural productivity on 5 million hectares of farmland since the 1990s and restored at least 250,000 hectares of severely degraded land by combining nitrogen-fixating trees with crops in agroforestry systems. Biodiversity in the region has risen, soil fertility has improved, and landscapes have become more resilient to water-related shocks. In addition, real farm incomes have doubled, with the gross annual income of the region increasing by 1,000 USD per household—all through nature-based solutions.

Another promising example of sustainable innovation comes from India, where encroachment on tiger habitat by villagers seeking firewood has threatened the survival of the big cats. To reduce such encroachment, hundreds to thousands of small biogas-fueled units have been installed in rural villages to enable families to cook on stoves using methane rather than firewood. When filled with 40 kg (88 pounds) or so of cattle dung and 40 liters (10 gallons) of water, a typical biogas unit can produce enough methane to cook three meals a day for a family of six. In some parts of northern India, where biogas units have been installed, fuel wood consumption has been reduced by 70 percent. In addition, as farmers have adopted stall feeding of cattle, in order to collect the dung, there have been fewer cases of tigers preying on the cattle. By this simple change in energy-usage patterns pressure has been reduced on the remaining few forests in the area, rendering hope for the small surviving population of tigers dependent on this resource for their survival.

We believe that a key reason why we don't see more of these extraordinarily efficient, sustainable, and attractive nature-based solutions go to scale isn't because we lack evidence that they work. Rather, it's because of the perverse set of incentives that we live with, and the lack of clear regulation. We live in a world where it makes economic sense to be inefficient in the way that we use natural resources (such as phosphorus in agriculture), ecosystems (by overfishing our oceans and cutting down our forests for short-term benefits for a few while undermining their value for the majority and for the future), and our atmosphere (through air pollution and climate change). In the short term, such behavior creates an illusion of success, because it feels like we can erode natural capital and emit greenhouse gases for free, with no planetary bill to pay. But in the long term we're all losers, as Earth's long-neglected bills come due in the form of droughts, diseases, ecosystem collapses, or extreme weather events. As long as we continue to undermine the ecosystems upon which we depend, we're taking a risky, unhealthy, and inefficient path into the future.

Correcting this massive global market failure is an urgent necessity. By calculating the true cost of all forms of pollution and planetary abuse, and by establishing regulations that enable economic development within planetary boundaries, we can protect Earth's remaining natural ecosystems *without* putting a lid on development. On the contrary, such measures will unleash innovation by making untapped, sustainable, nature-based solutions worthwhile to invest in. Instead of being a "limit" to growth, defining a safe operating space on Earth with absolute budgets for carbon, water, and land, will do just the opposite. It will unleash innovation, enabling us to achieve growth within limits for human prosperity in a "good" Anthropocene.

Hikers cross a ridge in Landmannalaugar, Iceland. Now that humanity has become the largest force of change on Earth, we are truly a big world on a small planet. Until now Earth has proven to be remarkably resilient to our impacts. Let's continue to safeguard nature so that the grandness of biomes like this vast glacier continue to support our wellbeing.

AFTERWORD
A NEW PLAYING FIELD

IN THIS BOOK we've described a new science-based approach to defining a safe operating space for humanity—an approach anchored by the planetary boundaries framework. The key to future prosperity, we've argued, is to safeguard the remaining beauty on Earth by reducing humanity's heavy footprint. No matter where we live, or what our walk of life might be, we all depend on stable, resilient natural systems such as the atmosphere, the oceans, and terrestrial ecosystems. To avoid triggering dangerous tipping points as the Earth responds to humanity's massive impacts, we must find ways to pursue economic growth, food security, and thriving communities in the years ahead without undermining the life forms that make our own lives possible and good.

The time has come for a change. World leaders such Jim Yong Kim, president of the World Bank, Ban Ki-moon, UN Secretary General, Christine Lagarde, managing director of the International Monetary Fund, and Angel Gurria, secretary general of the OECD, have all recognized the fact that allowing business-as-usual to continue will only endanger the world's prospects for growth and poverty alleviation. Corporate leaders, too, have increasingly adopted the idea that sustainability also happens to be good business. As Ulrich Spiesshofer, CEO of the Swiss industrial giant ABB, recently put it, "we need to run the world without consuming the Earth."

Sustainability, after all, isn't limiting. In fact, it encourages innovation in the same way that the lines on a soccer field make Lionel Messi's brilliance possible. If you know where the boundaries are, you can be a virtuoso of economic growth as creative as Zlatan Ibrahimovic is with a soccer ball. By defining a safe operating space, we can both preserve the natural world and pursue our own prosperity at the same time.

We've concluded that, for a long time, humanity got it all wrong. For centuries, we clung to the belief that we could have *growth without limits* on a finite planet. Then 40 years ago this belief collided with environmental arguments proposing *limits to growth*. We thought that as long as we kept our own backyards clean through regulations on chemicals, local air and water quality, and protection of ecosystems, we could have "sustainable development." Alas, how wrong we were. Earth proved to be much more complex than that. Our local environmental abuses had long-distance impacts. Because of the biophysical processes that link the polar regions with the savannahs, rainfall systems across the world, the oceans and atmosphere with local weather systems, what you and I hid under the carpet in one corner of the world came back to haunt someone else in another corner, often in the most abrupt and surprising ways. We never expected to destabilize the Greenland ice sheet, West Antarctic glaciers, tropical coral reefs, or the Siberian tundra simply by the way we ran our local economies.

So today we're proposing a new playing field of *growth within limits*. By combining what we know from science about Earth's biophysical limits with emerging evidence about transformative technologies and values, we see unlimited opportunities for abundance through a combination of wisdom, innovation, and worldwide collaboration. Our idea is to redefine sustainable development as *the pursuit of good lives for all within a safe and just operating space on Earth*. To guide us in this direction, we're offering an easy-to-remember but scientifically robust number—a number we can teach to our children, since they're the ones likely to be alive in the year 2100. That number is zero.

We know with a high degree of certainty that by the second half of this century we must reach a zero-emission global society as part of a fully decarbonized world economy. We must also attain a zero rate of species loss to halt declining biodiversity. Finally, having transformed half of the Earth's surface to farmland and cities, we must now find ways to feed the world within existing agricultural lands. Enough is enough, and in order to secure future rainfall, carbon sinks, and habitats for all living species, we must feed humanity with zero expansion of agricultural territory. The next green revolution will be truly green, occurring through sustainable intensification on current agricultural land.

This triple zero formula—zero emissions, zero loss of biodiversity, and zero expansion of agricultural land—constitutes a science-based agenda for world development that defines a significant part of Earth's safe operating space. It's an easy number to remember. A number with little uncertainty. A visionary goal to unleash a Second Machine Age that is green, resilient, and prosperous.

Why are we hesitating? Beats us.

Making the transition to a thriving world within a safe and just operating space has become not only necessary, but also possible and desirable. Based on the latest science, we've defined planetary boundaries to serve as positive guides for action and as incentives for exponential technologies. People all over the world, we believe, share a universal wish to cherish the natural world. All we need to do to achieve a safe and prosperous future is to become dedicated stewards of the remaining beauty on Earth.

FURTHER INFORMATION ON PHOTOGRAPHS

Page 15

Tebaran, a hunter in Borneo, sees a difficult future for indigenous people as logging operations engulf the rainforest that was the land of his ancestors. Tropical rainforests are also critical biomes that regulate Earth's resilience. Cutting down too much of a rainforest, combined with the warmer and often drier environment that climate change is expected to bring in the tropics, may cause an abrupt tipping point in the ecosystem, turning the rainforest into a savannah. If that were to happen, it could have damaging impacts on local freshwater supplies and trigger a massive loss of carbon.

Page 16

Thanks to a boom in construction projects like this one in Hong Kong, cities around the world continue to expand. According to projections, two thirds of the cities we will need by 2030 have not been built yet. This is a gigantic challenge but also a grand opportunity. We now know how to build cities that integrate ecosystems into their design to make them attractive, resilient, and healthy. Nature plays a key role in the design of sustainable cities.

Page 38

A pair of putty-nosed guenon, tree-dwelling monkeys, are offered for sale as bushmeat by poachers along a road in Cameroon. European fisheries policy may have unintentionally stimulated an increase in bushmeat hunting. When foreign fishing fleets were pushed away from European waters by new regulations, the big ships moved off the coast of Africa, where they cleaned out fish resources. That put small-scale African fishermen out of business. To make ends meet for their families, some of these fishermen became bushmeat hunters. The killing of these animals not only threatens biodiversity in the region, it also increases the risk of zoonotic disease outbreaks, such as happened with Ebola.

Page 80

When lakes and rivers are loaded with chemicals from urban and agricultural runoff, algae blooms can choke the water. Under such circumstances, the water can become so oxygen-deprived that fish cannot live in it. As we have increasingly discovered, safeguarding rich fish populations is critical to enabling marine environments such as lakes and coral reef systems to bounce back after environmental shocks. Following a major bleaching event, for example, a coral reef system can break down. But the presence of many varieties of grazing fish, such as parrot and surgeon fish, can help the reef regenerate itself. Without these aquatic "lawnmowers," a collapsed coral reef will be taken over by seaweed. If the system, moreover, is loaded with nutrients from agricultural runoff, it is even more likely to get stuck in a new murky, algae-dominated state.

Pages 122–123

Mangrove trees along the coast of West Papua provide nursery habitat for fish in their root systems. As shown recently by The Economics of Ecosystems and Biodiversity (TEEB) initiative, mangrove systems provide significant incomes for small-scale coastal communities if kept intact. By contrast, if sudden extreme flooding events degrade mangroves, the resulting loss of livelihoods can have devastating social costs.

Page 146

What does the future hold for a boy in Rwanda? Sustainable solutions offer the best chances to alleviate poverty. The key challenge in exploring sustainable food systems within planetary boundaries is to recognize that the era of farmland expanding into natural ecosystems has come to an end. We must now feed humanity on existing farmland, which will require sustainable intensification. This, in turn, will necessitate major innovations, combining cutting-edge science with indigenous knowledge.

Page 162

Orderly ranks of oil palms have replaced natural habitats for plants and animals in Sarawak. Since the 1980s and 1990s, forests in this part of Borneo have been leveled at an unparalleled rate. More tropical wood has been exported from Sarawak and the rest of Borneo during the past two decades than from Africa and South America combined. As logging concessions have expired, the land in many places has been converted to palm oil plantations, encouraged by subsidies from the government. Since the 1990s there has been a 10-fold increase in planned plantations. The largest palm oil plantation in the world, encompassing 800,000 hectares, has now been proposed on Borneo.

Page 180

A rapidly changing environment puts these social–ecological systems at risk. At the same time they can teach us a lot about how to work with nature for human wellbeing and resilience.

Previous pages: A vibrant rainforest surrounds Iguazu Falls, a series of cataracts on the border between Brazil and Argentina.

KEY SOURCES AND RECOMMENDED READING

Chapter 1: Our New Predicament

Friedman, T.L., 2005, *The World Is Flat: A Brief History of the Twenty-first Century*, Farrar, Straus, and Giroux, New York, p 475.

Gunderson, L. and Holling, C.S. (eds), 2002, *Panarchy: Understanding Trans-formations in Human and Natural Systems*, Island Press, Washington, DC.

Hansen, J.E., and Sato, M., 2012, *Paleoclimate Implications for Human-Made Climate Change*, Springer, Berlin, Germany.

Holling, C.S., 1973, "Resilience and Stability of Ecological Systems," *Annual Review of Ecology and Systematics*, 4: 1–23.

IPCC, 2014, http://www.ipcc.ch/

IUCN Redlist, 2014, http://www.iucnredlist.org/

MA, 2005, *Millennium Ecosystem Assessment: Ecosystems and Human Well-being: Synthesis,* Island Press, Washington DC.

NOAA National Weather Service, 2014, http://www.noaa.gov

OECD. 2014. "Economic outlook" http://www.oecd.org/eco/economicoutlook.htm

Oppenheimer, S. 2004, *Out of Eden: The Peopling of the World*, Constable & Robinson, London, UK. p 429.

Rockström, J.; Steffen, W.; Noone, K.; Persson, Å.; Chapin, III, F.S.; Lambin, E.F.; Lenton, T.M.; Scheffer, M.; Folke, C.; Schellnhuber, H.J.; Nykvist, B.; de Wit, C.A.; Hughes, T.; van der Leeuw, S.; Rodhe, H.; Sörlin, S.; Snyder, P.K.; Costanza, R.; Svedin, U.; Falkenmark, M.; Karlberg, L.; Corell, R.W.; Fabry, V.J.; Hansen, J.; Walker, B.; Liverman, D.; Richardson, K.; Crutzen, P.; and Foley, J.A. 2009, "A Safe Operating Space for Humanity," *Nature*, 461: 472–475.

Scheffer, M.; Carpenter, S.R.; Foley, J.A; Folke, C; and Walker, B. 2001, "Catastrophic Shifts in Ecosystems," *Nature* 413: 591–596.

Steffen W., et al. 2004, *Global Change and the Earth System: a Planet Under Pressure*, The IGBP book series, Springer, Berlin, Germany.

TEEB, 2010, *The Economics of Ecosystems and Biodiversity: Mainstreaming the Economics of Nature: A Synthesis of the Approach, Conclusions, and Recommendations of TEEB*, Progress Press, Malta.

UN DESA, 2014, http://www.un.org/en/development/desa/population/

Welcome to the Anthropocene, 2014, http:// www.anthropocene.info/en/home

Wilson, E.O., 2013, *The Social Conquest of Earth*, Liveright Publishing Corporation, New York, USA, p 327.

Young, O., and Steffen, W., 2009, "The Earth System: Sustaining Planetary Life Support Systems," in *Principles of Ecosystem Stewardship: Resilience-based Resource Natural Resource Management in a Changing World*, Chapin III, F.S; Kofinas, G.P; and Folke, C.; (eds.), pp 295. Springer, New York.

Chapter 2: Planetary Boundaries

AMAP, "AMAP Assessment 2013: Arctic Ocean Acidification," http://www.amap.no/documents/doc/AMAP-Assessment-2013-Arctic-Ocean-Acidification/881

Canadell, J.G.; Le Quéré, D.; Raupach, M.R.; Field, C.R.; Buitenuis, E.; Ciais, P.; Conway, T.J.; Gillett, N.P.; Houghton, R.A.; and Marland, G., 2007, "Contributions to Accelerating Atmospheric CO_2 Growth from Economic Activity, Carbon Intensity, and Efficiency of Natural Sinks", *Proceedings of the National Academy of Sciences*, 104: 18866–18870.

Carson, R., 2002, *Silent Spring: The Classic that Launched the Environmental Movement*, A Mariner Book Houghton Mifflin Company, Boston, USA.

Global Biodiversity Outlook 4. http://www.cbd.int/gbo4/

IGBP, IOC, SCOR, 2013, *Ocean Acidification: Summary for Policymakers – Third Symposium on the Ocean in a High-CO_2 World*, International Geosphere-Biosphere Programme, Stockholm, Sweden. http://www.igbp.net/download/18.30566fc6 142425d6c91140a/1385975160621/OA_spm2-FULL-lorez.pdf

IPCC, 2014, http://www.ipcc.ch/

Meadows, D.H., Randers, J.; and Meadows, D.L., 2004, *Limits to Growth: The 30 Years Update*, Chelsea Green Publishing Company, USA. p 325.

National Climate Assessment, 2014, http://nca2014.globalchange.gov/downloads

Rockström, J.; Steffen, W.; Noone, K.; Persson, Å.; Chapin, III, F.S.; Lambin, E.F.; Lenton, T.M.; Scheffer, M.; Folke, C.; Schellnhuber, H.J.; Nykvist, B.; de Wit, C.A.; Hughes, T.; van der Leeuw, S.; Rodhe, H.; Sörlin, S.; Snyder, P.K.; Costanza, R.; Svedin, U.; Falkenmark, M.; Karlberg, L.; Corell, R.W.; Fabry, V.J.; Hansen, J.; Walker, B.; Liverman, D.; Richardson, K.; Crutzen, P.; and Foley, J.A., 2009, "A Safe Operating Space for Humanity," *Nature*, 461: 472–475.

Rockström, J.; Steffen, W.; Noone, K. Persson, Å.; Chapin, III, F.S.; Lambin, E.F.; Lenton, T.M.; Scheffer, M.; Folke, C.; Schellnhuber, H.J.; Nykvist, B.; de Wit, C.A.; Hughes, T.; van der Leeuw, S.; Rodhe, H.; Sörlin, S.; Snyder, P.K.; Costanza, R.; Svedin, U.; Falkenmark, M.; Karlberg, L.; Corell, R.W.; Fabry, V.J.; Hansen, J.; Walker, B.; Liverman, D.; Richardson, K.; Crutzen, P.; and Foley, J.A., 2009, "Planetary Boundaries: Exploring the Safe Operating Space for Humanity," *Ecology and Society*, 14 (2): 32.

Rockström, J., et al., 2014, *Water Resilience for Human Prosperity*, Cambridge University Press, UK, p 284.

Steffen, W.; Richardson, K.; Rockström, J.; Cornell, S.; Fetzer, I.; Bennett, E.M., Biggs, R.; Carpenter, S.R.; de Vries, W.; de Wit, C.A.; Folke, C.; Gerten, D.; Heinke, J.; Mace, G.M.; Persson, L.M.; Ramanathan, V.; Reyers, B.; and Sörlin, S., "Planetary Boundaries: Guiding Human Development on a Changing Planet," in Review, *Science*, December 2014.

WMO/UNEP Scientific Assessments of Ozone Depletion, http://www.esrl.noaa.gov/csd/assessments/ozone/

World Water Development Report, 2014, http://www.unwater.org/publications/publications-detail/en/c/218614/

Chapter 3: Big Whammies

Barnosky, A.D.; Matzke, N.; Tomiya, S.; Wogan, G.O.U.; Swartz, B.; Quental, T.B.; Marshall, C.; McGuire, J.L.; Lindsey, E.L.; Maguire, K.C.; Mersey, B.; and Ferrer, E.A., 2011, "Has the Earth's Sixth Mass Extinction Already Arrived?" *Nature* 471: 51–57.

Bellwood, D.R.; Hughes, T.P.; Folke, C.; and Nyström, M., 2004, "Confronting the Coral Reef Crisis," *Nature* 429.

Box, Jason, 2012, The Meltfactor Blog, "Greenland Ice Sheet Record Surface Melting Underway," http://www.meltfactor.org/blog/?p=556

Burke, L.; Reytar, K.; Spalding, M.; and Perry, A.L., 2011, *Reefs at Risk Revisited*,

Washington, D.C., World Resources Institute, The Nature Conservancy, WorldFish Center, International Coral Reef Action Network, UNEP World Conservation Monitoring Centre and Global Coral Reef Monitoring Network, p 114.

Gleason D.F. and Wellington, G.M., 1993, "Ultraviolet Radiation and Coral Bleaching," *Nature* 365: 836–838.

Global Biodiversity Outlook 4, http://www.cbd.int/gbo4/ http://www.globalcarbonproject.org/ carbonbudget/index.htm

Lenton, T.M.; Held, H.; Kriegler, E.; Hall, J.W.; Lucht, W.; Rahmstorf, S.; Schellnhuber, H.J., 2008, "Tipping Elements in the Earth's Climate System," *Proceedings of the National Academy of Science*, 105 (6) : 1786-1793.

NOAA, 2013, Arctic Report Card, http:// www.arctic-report.net/

Smith, J.B.; Schneider, S.H.; Oppenheimer, M.; Yohe, G.W.; Hare, W.; Mastrandrea, M.D.; Patwardhan, A.; Burton, I.; Corfee-Morlot, J.; Magadza, C.H.D.; Füssel, H-M.; Barrie Pittock, A.; Rahman, A.; Suarez, A.; and van Ypersele, J-P., 2009, "Assessing Dangerous Climate Change Through an Update of the Intergovernmental Panel on Climate Change (IPCC): Reasons for Concern," *PNAS* 106 (11).

WBGU, 2009, *Solving the Climate Dilemma: The Budget Approach*, Special Report, German Advisory Council on Global Change.

Chapter 4: Peak Everything

Cohen, D. 2007, "Earth Audit," *New Scientist* 194 (2605): 34–41.

Cordell, D.; Drangert, J-O.; and White, S., 2009, "The Story of Phosphorus: Global

Food Security and Food for Thought," *Global Environmental Change* 19: 292–305.

GEA, *Global Energy Assessment*, 2012, Global Energy Assessment, *Toward a Sustainable Future*, Cambridge University Press, UK.

IEA, 2010, World Energy Outlook 2010, International Energy Agency, Paris, France.

IEA, 2014, World Energy Outlook 2014, http://www.worldenergyoutlook.org/ publications/weo-2014/

Meadows, D.H.; Meadows, D.L.; and Randers, J., 1992, *Beyond the Limits*, Chelsea Green Publishing Co., White River Junction, VT, USA.

Meadows, D.H.; Meadows, D.L.; and Randers, J., 2004, *Limits to Growth: The Thirty Year Update*, Chelsea Green Publishing Co., White River Junction, VT, USA.

Ragnarsdottir, K.V., 2008, "Rare Metals Getting Rarer," *Nature Geoscience* 1 (11): 720–721.

Chapter 5: No Business on a Dead Planet

Biggs, R.; Schlüter, M.; Biggs, D.; Bohensky, E.L.; Burnsilver, S.; Cundill, G.; Dakos, V.; Daw, T.; Evans, L.; Kotschy, K.; Leitch, A.; Meek, C; Quinlan, A.; Raudsepp-Hearne, C.; Robards, M.; Schoon, M.L.; Schultz, L.; and West, P.C., 2012, "Towards Principles for Enhancing the Resilience of Ecosystem Services," *Annual Review of Environment and Resources* 37: 421–448.

Costanza R.; deGroot, R.; Sutton, P.; van der Ploeg, S.; Anderson, S.; Kubiszewski, I.; Farber, S.; and Turner, R.K., 2014, "Changes in the Global Value of Ecosystem Services," *Global Environmental Change* 26: 152–158.

Elmqvist, T., 2012, *Cities and Biodiversity*

Outlook: Action and Policy, UN Secretariat of the Convention of Biological Diversity, Montreal, CAN, 66pp http://cbobook.org/

MA, 2005, *Millennium Ecosystem Assessment: Ecosystems and Human Well-being: Synthesis*, Island Press, Washington, DC.

Stern, N., 2006, *Review on the Economics of Climate Change*, H.M. Treasury, UK, October. http://www.sternreview.org.uk.

TEEB, 2010, *The Economics of Ecosystems and Biodiversity: Ecological and Economic Foundations*, Edited by Pushpam Kumar. Earthscan, London.

The New Climate Economy, 2014, *Better Growth, Better Climate. The New Climate Economy Report*. The Global Report. The Global Commission on the Economy and Climate, www.newclimateeconomy.report

WBCSD, 2011, *Vision 2050: The New Agenda for Business*. The World Business Council for Sustainable Development (WBCSD), http://www.wbcsd.org/vision2050

Chapter 6: Unleashing Innovation

Bartolino, S.; Bonatti, L.; and Sarracino, F., 2014, "Great Recession and U.S. Consumers' Bulimia: Deep Causes and Possible Ways Out," *Cambridge Journal of Economics* 38(5): 1015–1042

Brynjolfsson, E., and McAfee, A., 2014, *The Second Machine Age: Work, Progress, and Prosperity in a Time of Brilliant Technologies*, W.W. Norton and Company, New York, USA.

Diamandis, P. and Kotler, S., 2012, *Abundance: The Future is Better Than You Think*, Free Press: New York.

Ellen MacArthur Foundation (Ed.) 2014, *A New Dynamic: Effective Business in a Circular Economy*. Ellen MacArthur Foundation Publishing, http://www.ellenmacarthurfoundation.org/

Elmqvist, T., 2012, *Cities and Biodiversity Outlook: Action and Policy*, UN Secretariat of the Convention of Biological Diversity, Montreal, 66pp, http://cbobook.org/

Foley, J.A.; Ramankutty, N.; Brauman, K.A.; Cassidy, E.S.; Gerber, J.S.; Johnston, M; Mueller, N.D.; O'Connell, C.; Ray, D.K.; West, P.C.; Balzer, C.; Bennett, E.M.; Carpenter, S.R.; Hill, J.; Monfreda, C.; Polasky, S.; Rockström, J.; Sheehan, J.; Siebert, S.; Tilman, D.; and Zaks, D.P.M., 2011, "Solutions for a Cultivated Planet," *Nature* 478 (7369): 337–342.

GEA, Global Energy Assessment, 2012, *Global Energy Assessment: Toward a Sustainable Future*, Cambridge University Press, UK, http://www.naturalstep.org/

von Weizsäcker, E., Hargroves, K., Smith, M., Desha, C., and Stasinopoulos, P., 2009, Factor 5: *Transforming the Global Economy through 80% Increase in Resource Productivity*, Earthscan, London.

Chapter 7: Rethinking Stewardship

Griggs, D.; Stafford-Smith, M.; Gaffney, O.; Rockström, J.; Ohman, M.C., Shyamsundar, P.; Steffen, W.; Glaser, G.; Kanie, N.; and Noble, I., 2013, "Sustainable Development Goals for People and Planet," *Nature* 495(7441): 305–307.

Olsson, P., Folke, C., and Hughes, T.P., 2008, "Navigating the Transition to Ecosystem-based Management of the Great Barrier Reef," Australia. *Proceedings of the National Academy of Sciences* 105: 9489–9494.

PBL, 2009, *Getting into the Right Lane for 2050: A primer for EU Debate*, Netherlands Environmental Assessment Agency and Stockholm Resilience Centre. PBL, Bilthoven, the Netherlands, p 106.

Raworth, K., "A Safe and Just Space for Humanity: Can We Live Within the Doughnut?" Oxfam Discussion Paper (Oxfam, 2012)

SHELL, 2008, *Shell Energy Scenarios to 2050*, SHELL, the Hague, the Netherlands, p 5.

UN Open Working Group proposal for Sustainable Development Goals. http://sustainabledevelopment.un.org/focussdgs.html

Walker, B.H. and Salt, D., 2006, *Resilience Thinking: Sustaining Ecosystems and People in a Changing World*, Island Press, Washington, DC.

Chapter 8: A Dual-Track Strategy

Folke, C., and Rockström, J., 2011, 3rd Nobel Laureate Symposium on Global Sustainability: "Transforming the World in an Era of Global Change," Guest Editorial, *Ambio*, 40(7): 717–718.

Comprehensive Assessment of Water Management in Agriculture (CA), 2007, *Water for Food, Water for Life: A Comprehensive Assessment of Water Management in Agriculture*, London: Earthscan and Colombo: International Water Management Institute

GEA, Global Energy Assessment, 2012, *Global Energy Assessment: Toward a Sustainable Future. Cambridge,* Cambridge University Press, *http://newclimateeconomy.report/*

Rockström, J. and Falkenmark, M., 2000, "Semiarid Crop Production from a Hydrological Perspective: Gap Between Potential and Actual Yields," *Critical Reviews in Plant Sciences* 19 (4): 319–346.

UNEP, 2011, *Towards a Green Economy: Pathways to Sustainable Development and Poverty Eradication*, www.unep.org/greeneconomy

WBGU, 2011, *World in Transition – A Social Contract for Sustainability*. Flagship Report 2011. German Advisory Council on Global Change.

Chapter 9: Solutions from Nature

Barron, J.; Enfors, E.; Cambridge, H.; and Adamou, M., 2010, "Coping with Rainfall Variability: Dryspell Mitigation and Implication on Landscape Water Balances in Small-scale Farming Systems in Semi-Arid Niger," *International Journal of Water Resources Development* 26: 523–542.

Benyus, J.M., 1997, *Biomimicry: Innovation Inspired by Nature*, Morrow, New York.

Lind, F. and Källström, N., 2014, "The Economic Case for Revitalizing the Baltic Sea," Boston Consulting Group March 2014. https://www.bcgperspectives. com/content/articles/corporate_social_responsibility_commu- nity_economic_development_economic_ case_revitalizing_baltic_sea/

Folke, C.; Jansson, Å.; Rockström, J.; Olsson, P.; Carpenter, S.R.; Chapin III, F.S.; Crépin, A-S.; Daily, G.; Danell, K.; Ebbesson, J.; Elmqvist, T.; Galaz, V.; Moberg, F.; Nilsson, M.; Österblom, H.; Ostrom, E.; Persson, Å.; Peterson, G.; Polasky, S.; Steffen, W.; Walker, B.; and Westley, F., 2011, "Reconnecting to the Biosphere," *Ambio* 40 (7): 719–738.

Hawken, P.; Lovins, A.B.; and Lovins, L.H., 2005, *Natural Capitalism: the Next Industrial Revolution*. 2nd Ed., Routledge, London.

IAASTD, 2009, *Agriculture at a Crossroads: International Assessment of Agricultural knowledge, Science and Technology for Development* (IAASTD), Summary for decision-makers of the Global Report.

Jackson, T., 2009, *Prosperity without Growth: Economics for a Finite Planet*, Earthscan, London.

Olsson, P., and V. Galaz., 2012,
"Social-ecological Innovation and
Transformation," in: Nicholls, A. and
Murdoch, A. (eds), *Social Innovation:
Blurring Boundaries to Reconfigure
Markets*, Palgrave MacMillan.

Pauli, G., 2010, *Blue Economy*, Paradigm
Publications.

The New Climate Economy, 2014, *Better
Growth, Better Climate: The New Climate
Economy Report*. The Global Report. The
Global Commission on the Economy and
Climate. www.newclimateeconomy.report

Zoltan, T.; Imredy, J.P.; Bingham, J-P.;
Zhorov, B.S.; and Moczydlowski, E.G.,
2014, "Interaction of the BKCa Channel
Gating Ring with Dendrotoxins,"
Channels. Volume 8, Issue 5.

**Graphs, Illustrations, and Tables
(for full references, see section above)**

Figure 1.1 Steffen, et al., 2004, updated by
Steffen, W., Deutsch, L. et al., 2014.

Figure 1.3 Adapted from *National
Geographic* magazine, March 2011.

Figure 1.5 www.regimeshifts.org

Table pages 50–51 www.regimeshifts.org

Figure 1.6 Hansen et al., 2012.

Figure 1.7 Hansen et al., 2012.

Figure 1.8 Adapted from Young and Steffen,
2009.

Table page 61 Steffen, W., and others.
Forthcoming 2014.

Figure 2.1 Steffen, W., and others.
Forthcoming 2014.

Figure 2.2 Adapted from Canadell, et al.,
2007.

Figure 3.1 Adapted from Smith, et al., 2009
and IPCC, 2014.

Figure 3.3 Adapted from WBGU, 2009.

Figure 3.4 Adapted from Donner, S.D.,
2009, and the World Resources Institute
project report, *Reefs at Risk Revisited*,
2011.

Figure 4.1 Adapted from *New Scientist*,
2007.

Figure 4.2 Adapted from Kjell Aleklett, pers.
comm., and Worldwatch Institute, 2006.

Table page 119 Adapted from TEEB, 2010.

Figure 7.1 Adapted from original by Carl
Folke, Stockholm Resilience Centre.

Figure 7.2 Adapted from Raworth, 2012
(Oxfam).

Figure 8.1 Adapted from WBGU, 2011.

Figure 9.1. Adapted from the New Climate
Economy Report, 2014.

Johan Rockström is the Director of the Stockholm Resilience Centre and a Professor in Water Systems and Global Sustainability at Stockholm University. An internationally recognized scientist on global sustainability issues, he leads the science on planetary boundaries—an increasingly established approach to human prosperity in the Anthropocene, which is at the heart of *Big World Small Planet*. He is an advisor to several governments, international policy processes, and business networks, and has published several books and over 100 scientific articles. He chaired the design phase for Future Earth and currently chairs the Earth League, the EAT Initiative, and the CGIAR program on Water, Land, and Ecosystems. He was honored as "Swede of the Year" in 2009, and voted Sweden's most influential person on environmental issues in 2012 and 2013.

Mattias Klum is a freelance photographer and filmmaker from Sweden. In an artistic way that is entirely his own, he has, since 1986, described and portrayed endangered species, natural environments, and ethnic minorities in peril around the globe. Since 1997, he has contributed a number of articles and cover images for *National Geographic* magazine. Klum was named a Young Global Leader of the World Economic Forum in 2008. Klum has been appointed an Ambassador for IUCN and WWF and serves as a member of the Board of Trustees at WWF Sweden. Klum is a Senior Fellow at the Stockholm Resilience Centre, a Fellow at the National Geographic Society and the Linnean Society of London. In 2013 Klum was awarded an honorary Doctorate in Natural Sciences by Stockholm University. This is Klum's thirteenth book.

ACKNOWLEDGMENTS

This book is the result of an interdisciplinary team effort encompassing science, photography, and storytelling. Without the vast treasure of scientific insights from colleagues at the Stockholm Resilience Centre (SRC), and the immense photographic experience in the Tierra Grande team we would never have been able to distill the science-based photographic story that resulted in this book. We are particularly indebted to Fredrik Moberg, head of Albaeco and senior communicator at the SRC for his research support, to Monika Klum and Fredrika Stelander at Tierra Grande for their help in selecting photographs and connecting them to the text. Great thanks to Anders Backlund, Maria Backlund, and Matilda Valman for your critical reviews and fact-checking. Jeppe Wikström, publisher at Max Ström, thanks for your engagement and strategic advice on how to configure an easy-to-read book about the world's greatest complexities, and Anna Sanner, project coordinator at Max Ström. We are grateful to Jerker Lokrantz for the design of all the graphics and to Patric Leo for the design of the book. *National Geographic* magazine and National Geographic Missions Program provided important inputs in the design of this book project, opening the door to an invaluable collaboration with Peter Miller, our editorial magician, who with the support of Postkodstiftelsen has shown the enviable skill of shortening, refining, and simplifying complex scientific text, without becoming simplistic. Finally, this book project was made possible by the generous core support given to the SRC by the Mistra Foundation.